French Aces of WWI

Volume 2: H–W

Alexandr Kazakov & Anton Zhukov

Acknowledgements

For our aviation books in print, please see our website at: **www.aeronautbooks.com**.
You may contact me at **jherris@me.com**

ISBN: **978-1-953201-13-3**

Design and layout: Jack Herris
Cover design: Aaron Weaver
Digital photo editing: Jack Herris

Aeronaut Books
Books for Enthusiasts by Enthusiasts
www.aeronautbooks.com

Table of Contents

P

R

S

T

V

W

Foreword

There are a huge number of publications devoted to the history of aviation of the First World War. There described in detail the course of military operations, the structure of the air forces of different countries, the biographies of pilots (especially aces), the history of various aviation units, as well as the material part and the history of the construction of warplanes of this period. Furthermore, a certain place is assigned to the colors and markings of the aircraft.

While working on the reference book *French aces of the First World War*, published in Russian in 2018, the authors of this album accomplished hundreds of color profiles of the planes on which the French aces of the First World War flew (including a small number that managed to fight during the Second World War). However, only 180 such profiles were included in the directory. So our task in this publication is to release about 500 color images of French aces' planes, many of which are published for the first time.

As a matter of course, the authors had to process an enormous aggregation of printed material such as books, magazines, and other periodicals over the many years. The basis for creating profiles was foremost photos, though some of them were made according to the narrative descriptions of eyewitnesses of the events of 1914–1918, although there were relatively few of them. A significant contribution was also made by Internet sites, where we managed to find a great deal of useful information. Finally, personal communication with serious historians on this topic has also greatly enriched this work. The book includes not only fighter planes on which aces gained the majority of their victories, but also military aircraft of other purposes, on which pilots began their combat careers, up to those on which they were trained in flying skills. Apart from the side images of aircraft, the book also includes some of their top-view projections when important marking elements were placed on the wings or on the top of the fuselage that could not be seen on the side projection. The book took the authors more than three years of painstaking work, and now this publication is the most complete collection of color profiles of French aces of the First World War, published in one edition.

We only hope that not only historians of the First World War and technology of that period will be interested in this work, but also modelers, manufacturers of models and aircraft decals in various scales, and a wide range of readers taking interest in various aspects of military and aviation history. So everyone could easily find a great deal of useful and really unique information.

Anton Zhukov and Alexander Kazakov

Claude Marcele Emile Haegelen

(1896–1950)

Sous-Lieutenant

F 8, N 3, Spa 100, Spa 89, Spa 48, SPA 162

22 victories

Spad 7
Spa 100
August 1918

Curtiss H-75 A-2 No. 183 (U-083)
Patroulle de DAT de Bourge
May–June 1940

Marius René Marie Hasdenteufel

(1894–1918)
Sous-Lieutenant
N 57, Spa 57
5 victories

SPAD 7
Spa 57
June 1918

Henry Joseph Marie Hay de Slade
(1893–1979)
Capitaine
N 80, N 86, Spa 159
19 victories

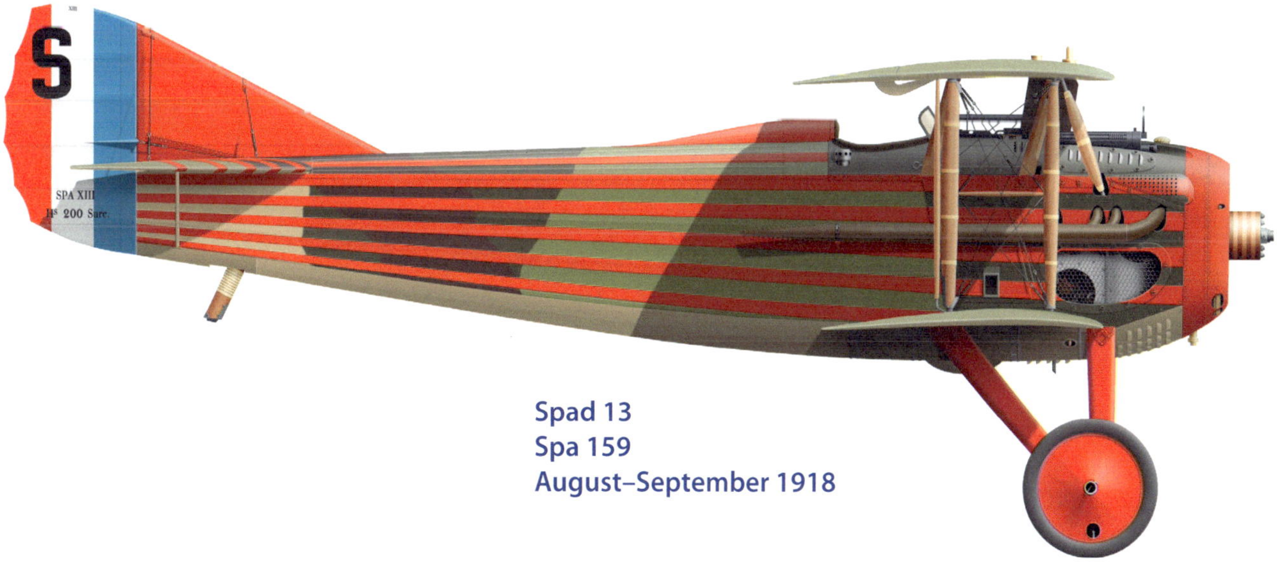

Spad 13
Spa 159
August–September 1918

Marcel Laurent Jean Baptist Henriot

(1886–1952)
Adjudant
N65, Spa65
6 victories

Spad 7 S1379
Spa 65
Summer 1917

Spad 7
Spa 65
February 1918

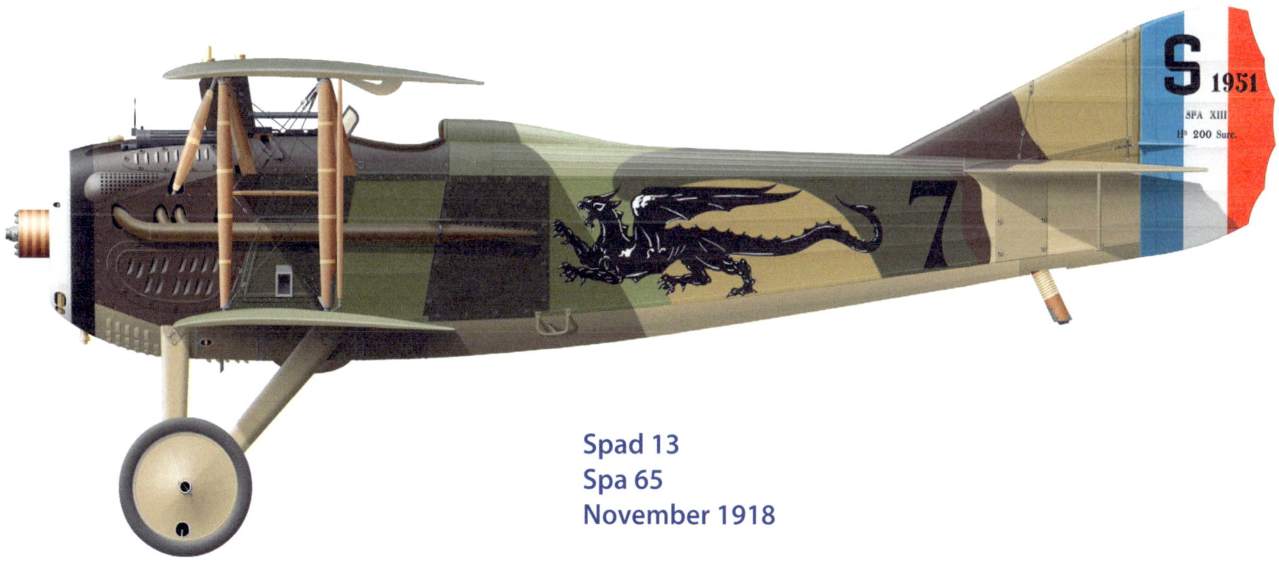

Spad 13
Spa 65
November 1918

Andre René Celestin Herbelin
(1889–1966)
Sous-Lieutenant
N 81, N 102, Spa 97
11 victories

Nieuport 12*bis Risque Tout*
N 102
Late 1916 – Early 1917

Nieuport 17 *Risque Tout*
N 102
Spring 1917

Spad 7 *Risque Tout VI* S1396
Spa 81
October 1917

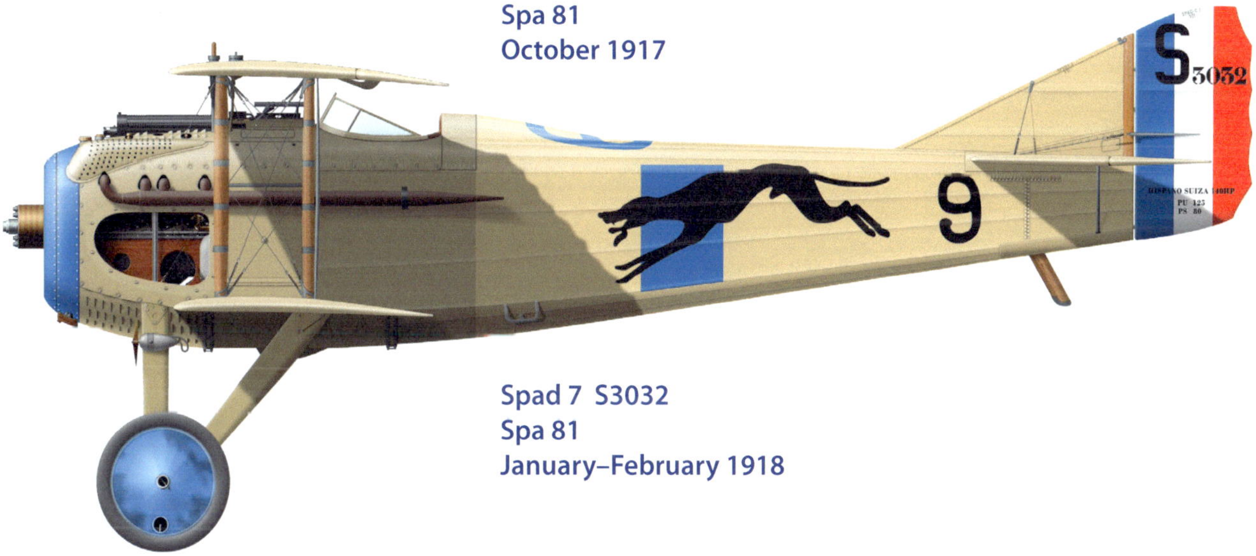

Spad 7 S3032
Spa 81
January–February 1918

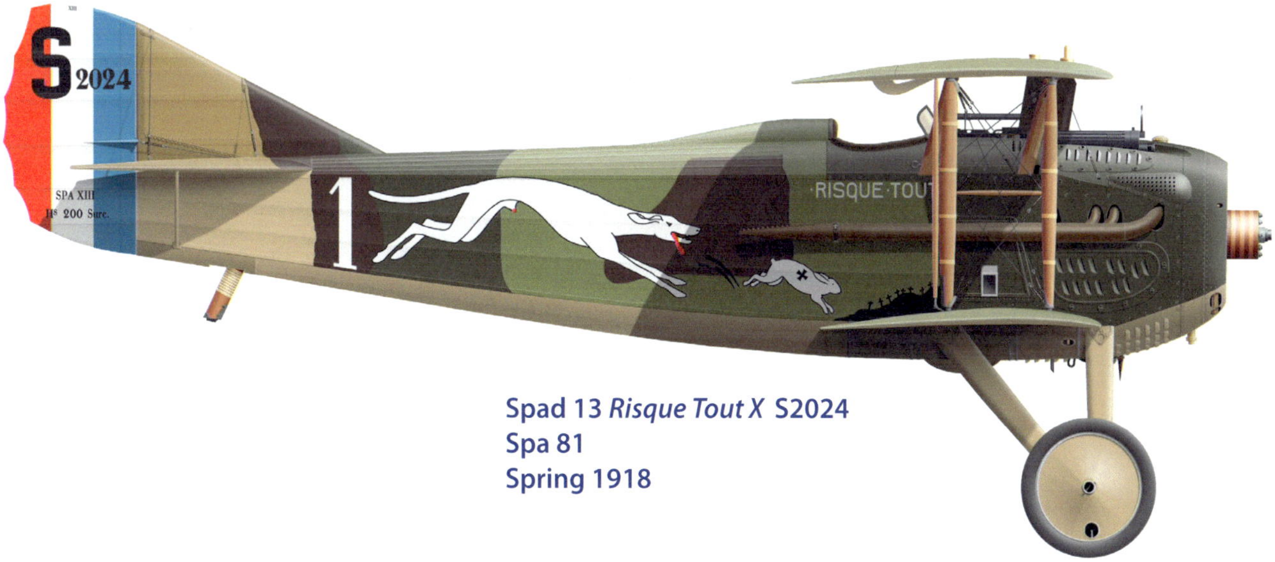

Spad 13 *Risque Tout X* S2024
Spa 81
Spring 1918

William Louis Max Herisson
(1894–1969)
Sous-Lieutenant
F 20, N 75, Spa 75
11 victories

Nieuport 24 N3987
N 75
Spring 1917

Spad 7 *Bousbous II*
Spa 75
Autumn 1917

Spad 12
Spa 75
Winter 1918

Alfred Marie Joseph Heurtaux
(1893–1985)
Capitaine
MS 26, MS 38, N 3, Spa 3
21 victories

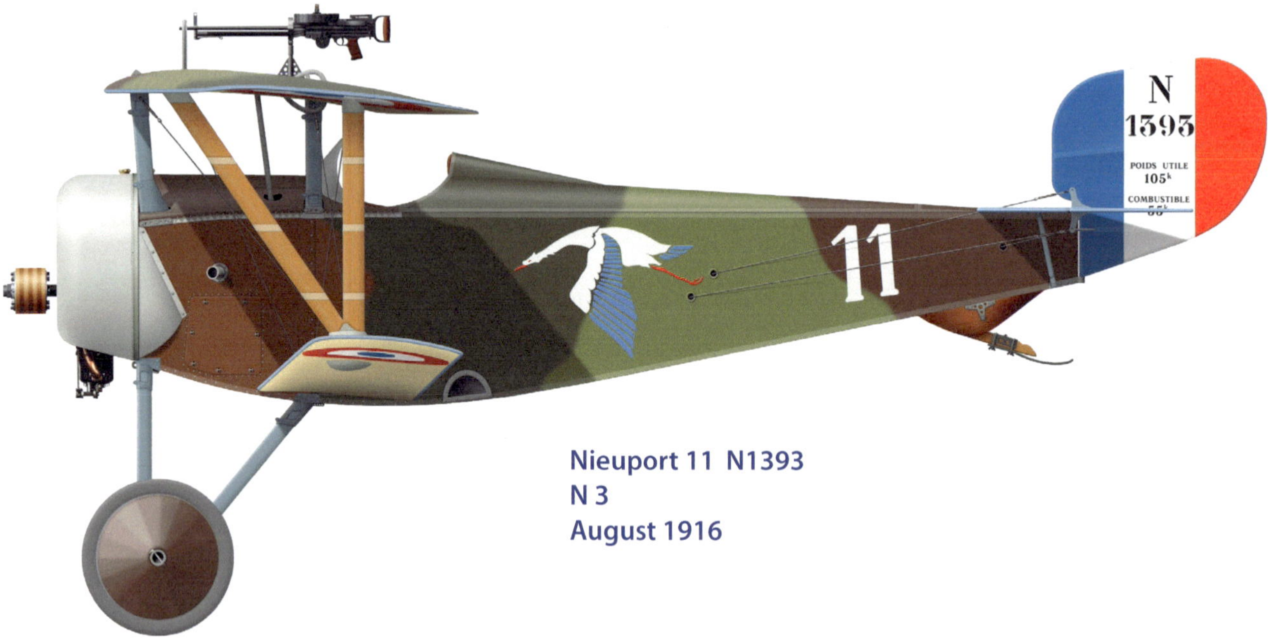

Nieuport 11 N1393
N 3
August 1916

Nieuport 11 N1393
N 3
September–October 1916

Nieuport 17 N1538
N 3
October 1916

Spad 7 S113
N 3
Autumn 1916

Paul Constant Homo
(1892–1968)
Capitaine
C 202, C 225, BR 235
5 victories

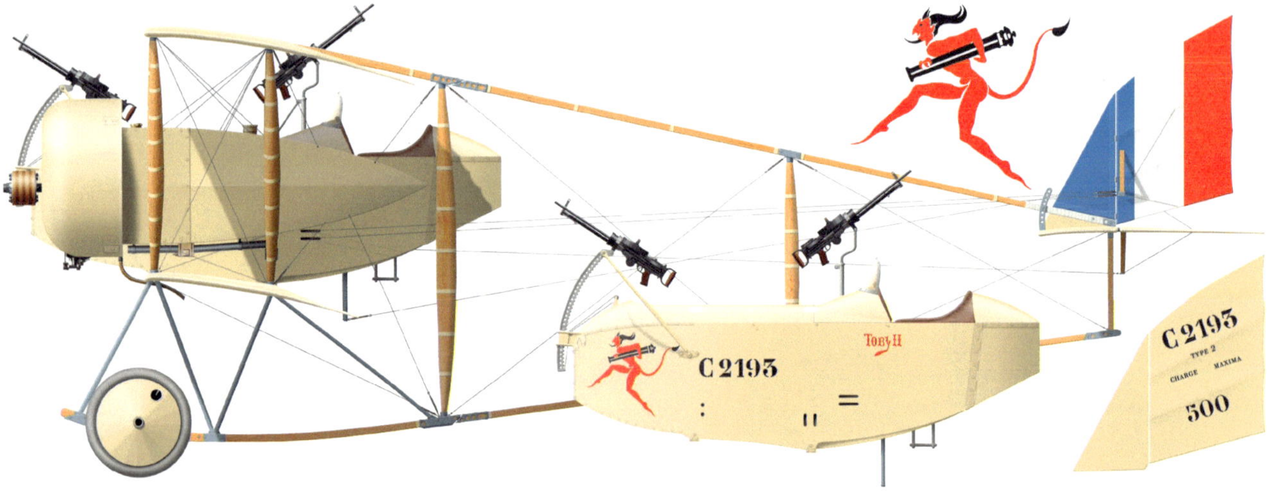

Caudron G.4 C.2193 *Toby II*
C 202
June 1916

Marcel Anatole Marie Esprit Hugues (1892–1982)

Capitaine
MF 22, N 77, N 81, Spa 95
12 victories

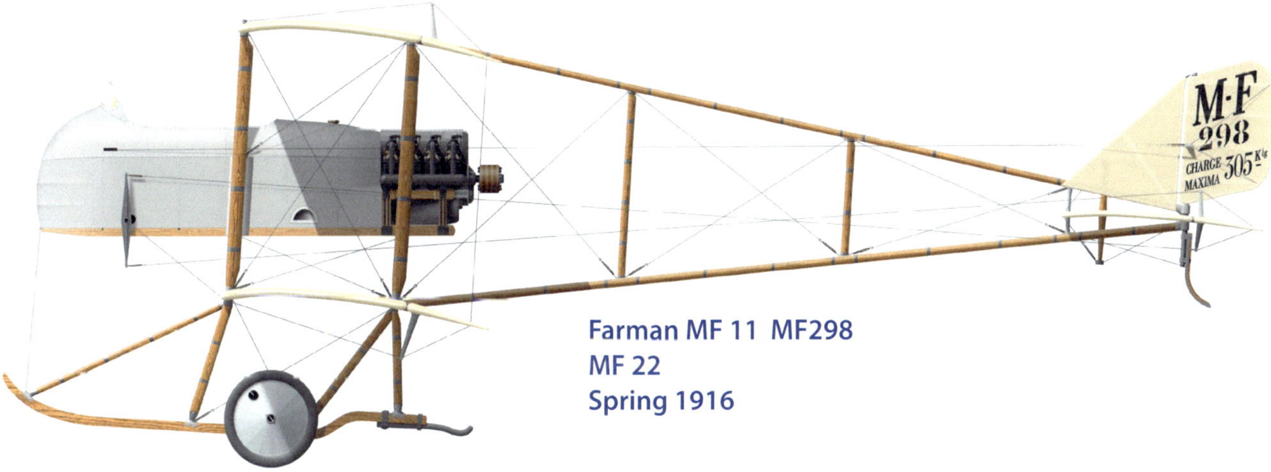

Farman MF 11 MF298
MF 22
Spring 1916

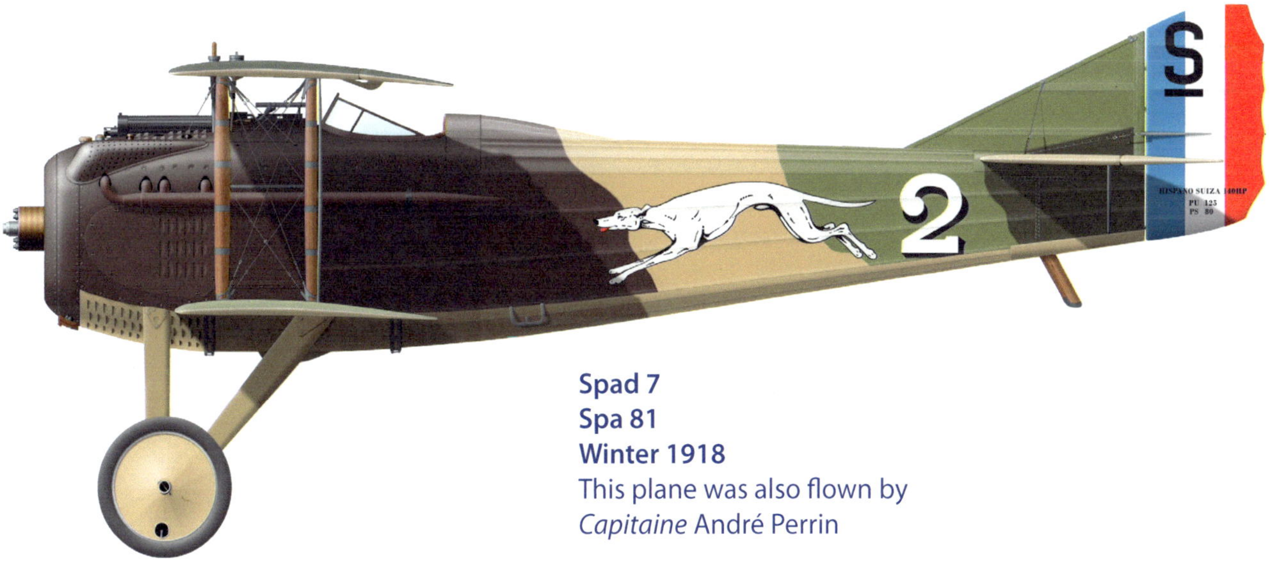

Spad 7
Spa 81
Winter 1918
This plane was also flown by *Capitaine* André Perrin

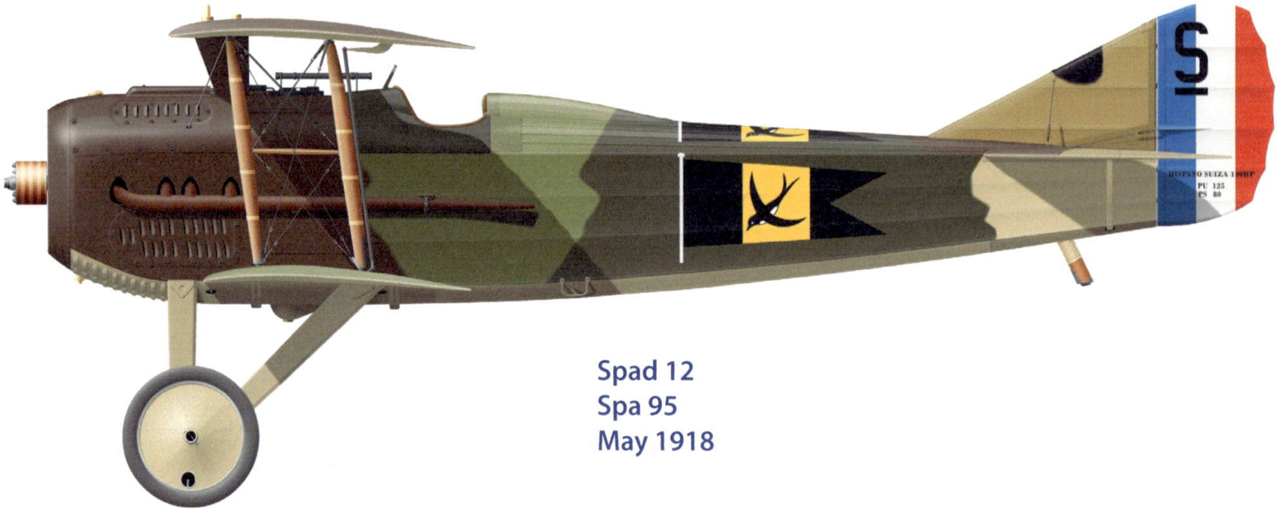

Spad 12
Spa 95
May 1918

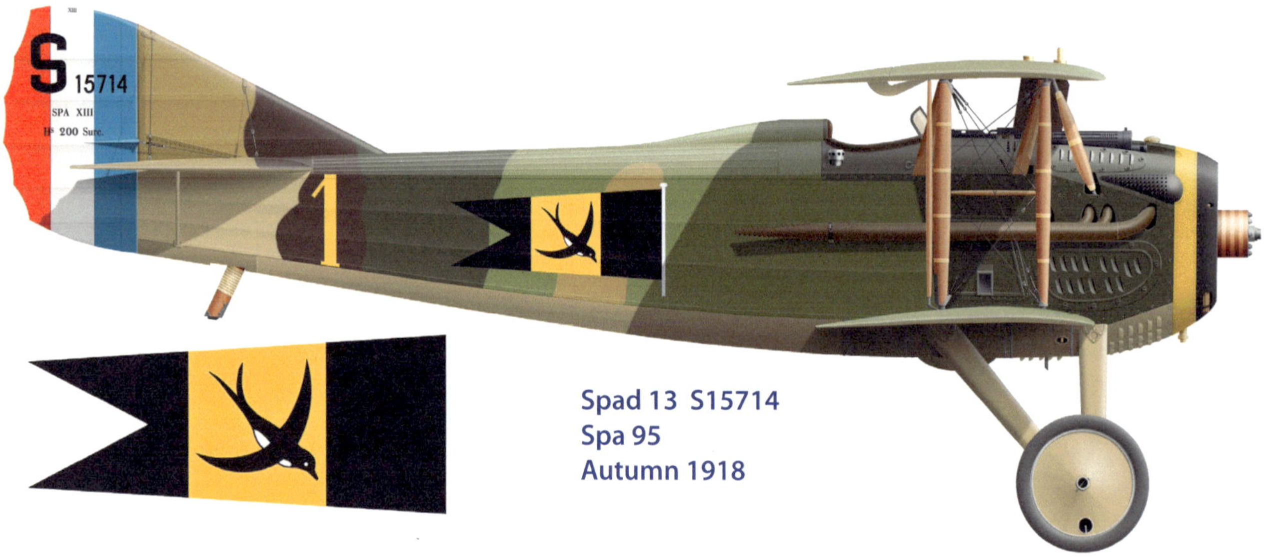

Spad 13 S15714
Spa 95
Autumn 1918

Curtiss H-75 A2 No.144 (U-044)
4 *Escadrille GC* II/5
May 1940

Lucien Joseph Jailler
(1889–1921)
Sous-Lieutenant
N 15, Spa 15
12 victories

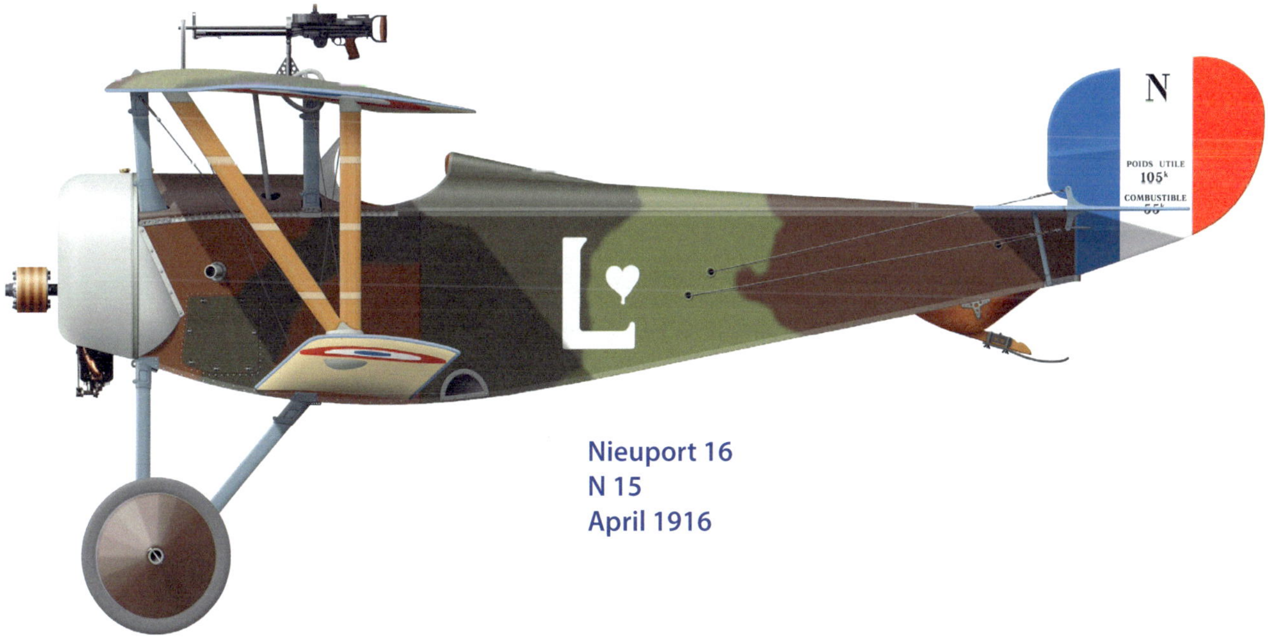

Nieuport 16
N 15
April 1916

Jean François Jannekeyn
(1892–1971)
Capitaine
Br 132
5 victories

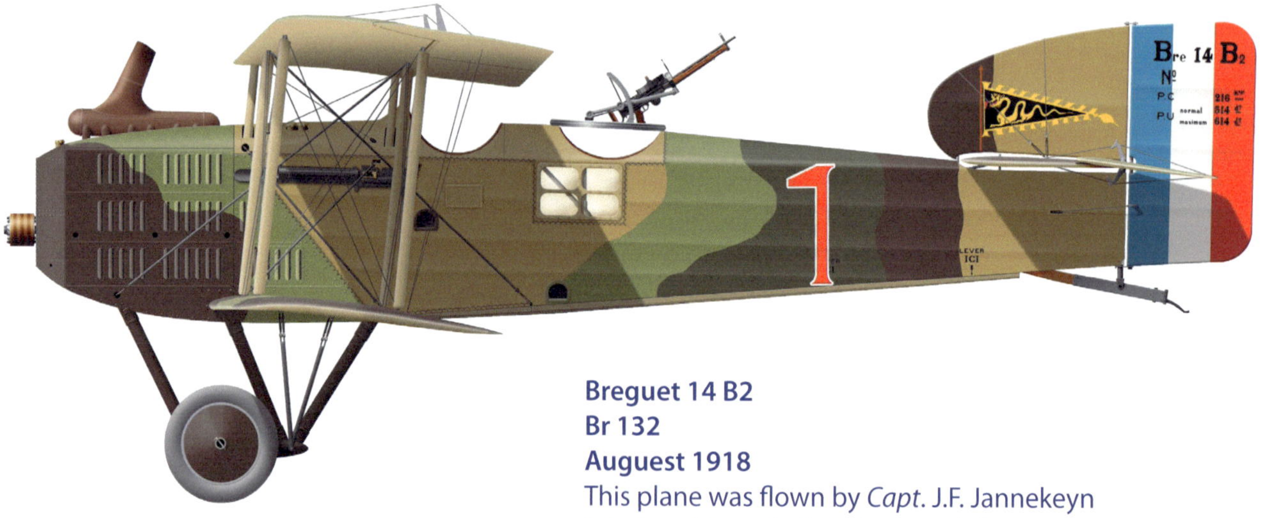

Breguet 14 B2
Br 132
Auguest 1918
This plane was flown by *Capt.* J.F. Jannekeyn (obs.) and *Sous-Lt.* E. Weismann (pilot)

Georges Marcel Lachmann
(1890–1961)
Capitaine
REP 27, REP 15, MS 15, N 57, N 92, N 581
9 victories

R.E.P. Type K (R54)
REP 27
January 1915
(Probably)

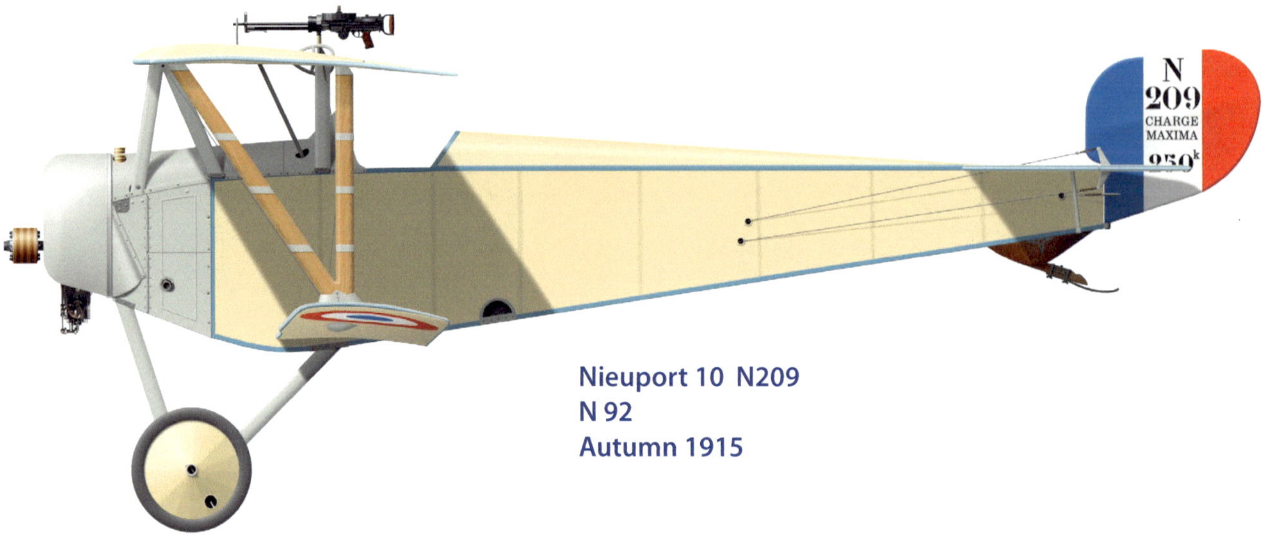

Nieuport 10 N209
N 92
Autumn 1915

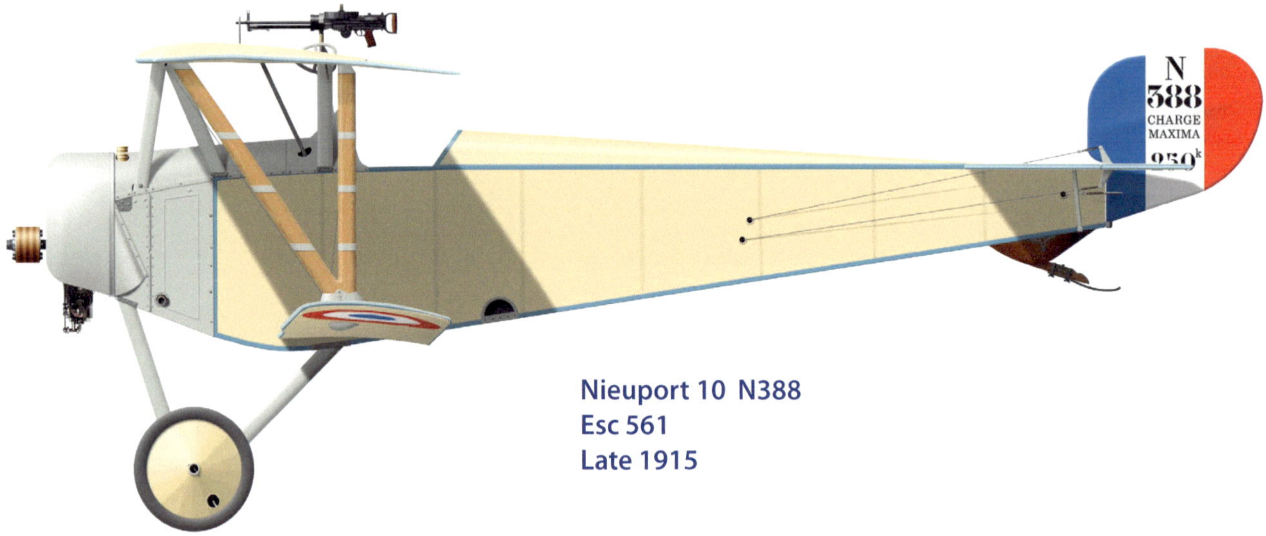

Nieuport 10 N388
Esc 561
Late 1915

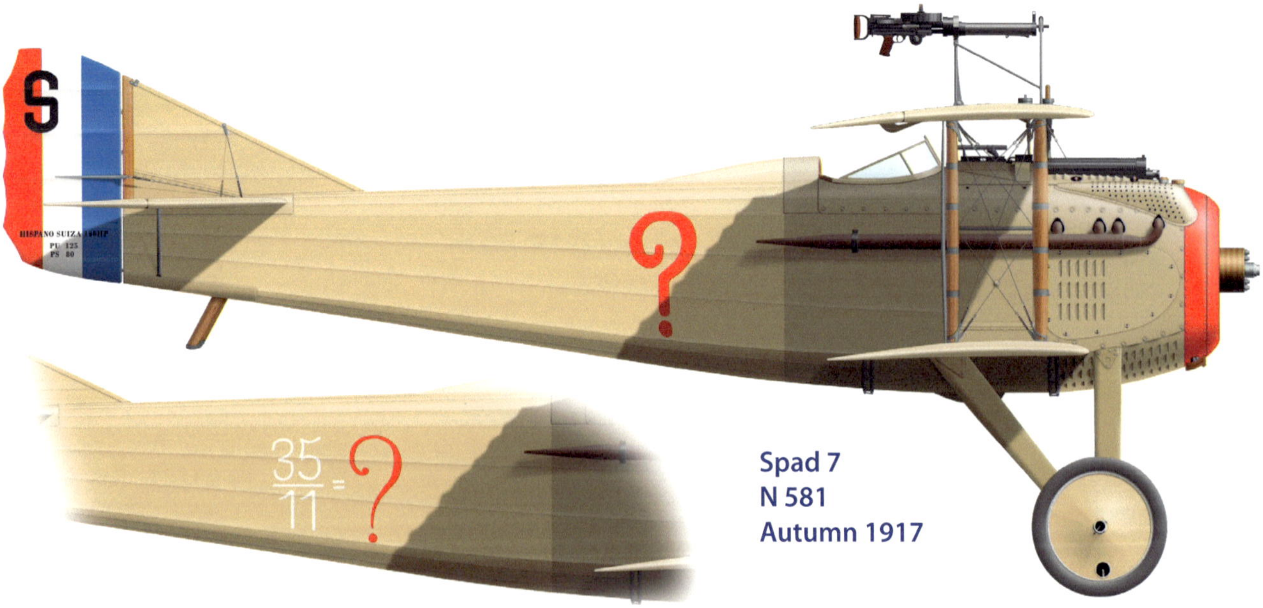

Spad 7
N 581
Autumn 1917

Spad 7
N 590
Autumn 1917

Spad 7 S11255
Br 590
June 1919
Lachmann fought against the
Hungarian revolution in this plane

Henri François Languedoc
(1885–1917)
Sous-Lieutenant
N 12
7 victories

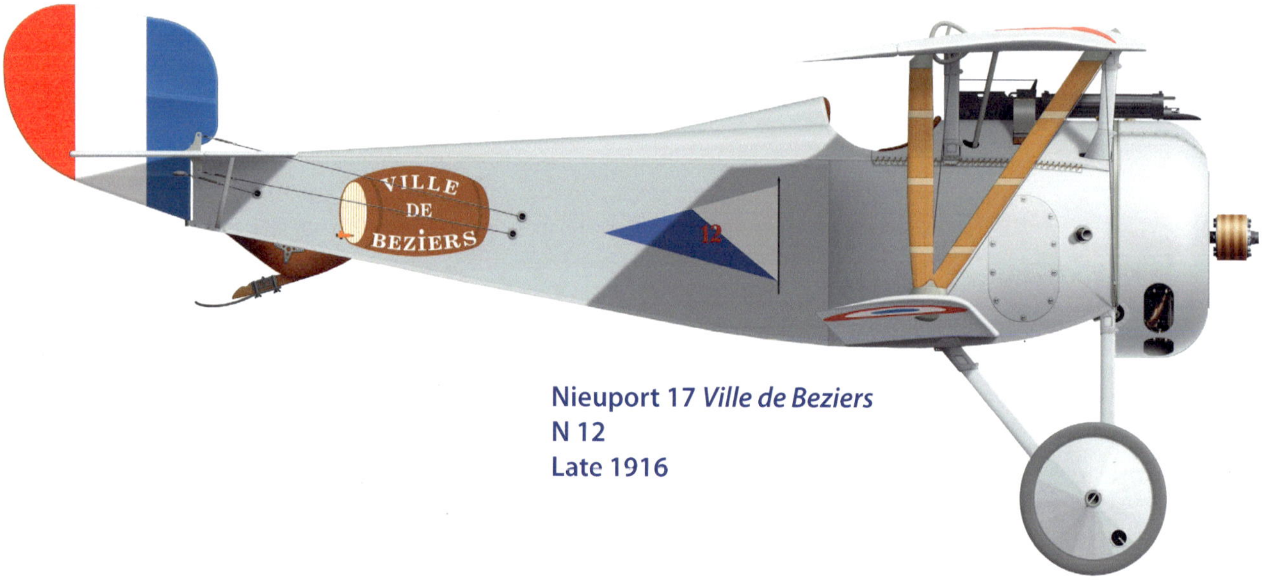

Nieuport 17 *Ville de Beziers*
N 12
Late 1916

Nieuport 17 *Lou CAMEL*
N 12
Early 1917

Antoine Laplasse
(1883–1918)
Adjudant-Chef
461CPR, Spa 75
8 victories

Nieuport 17
N 461 CPR
Summer 1917

Marie Charles Maurice Lecoq de Kerland
(1887–1978)
Lieutenant
N68, N90, Spa82
7 victories

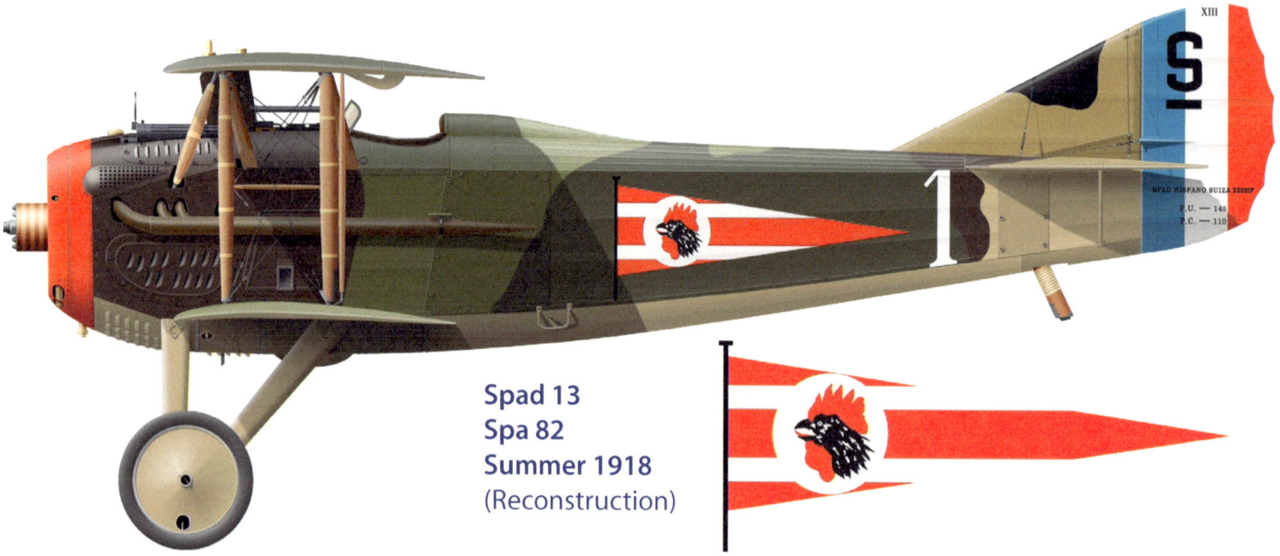

Spad 13
Spa 82
Summer 1918
(Reconstruction)

Didier Louis Marie Charles Lecour Grandmaison (1889–1917)

Capitaine
C 46, C 47
5 victories

Caudron R.4
C 46
September 1916

Letord 1
C 46
May 1917

Charles Eugene Joseph Marie Lefevre
(1893–1948)
Capitaine
N 15, N 152, Spa 96
6 victories

Nieuport 27 N5587
Spa 152
October 1917

Maxime Albert Lenoir
(1888–1916)
Adjudant
C 18, N 23
11 victories

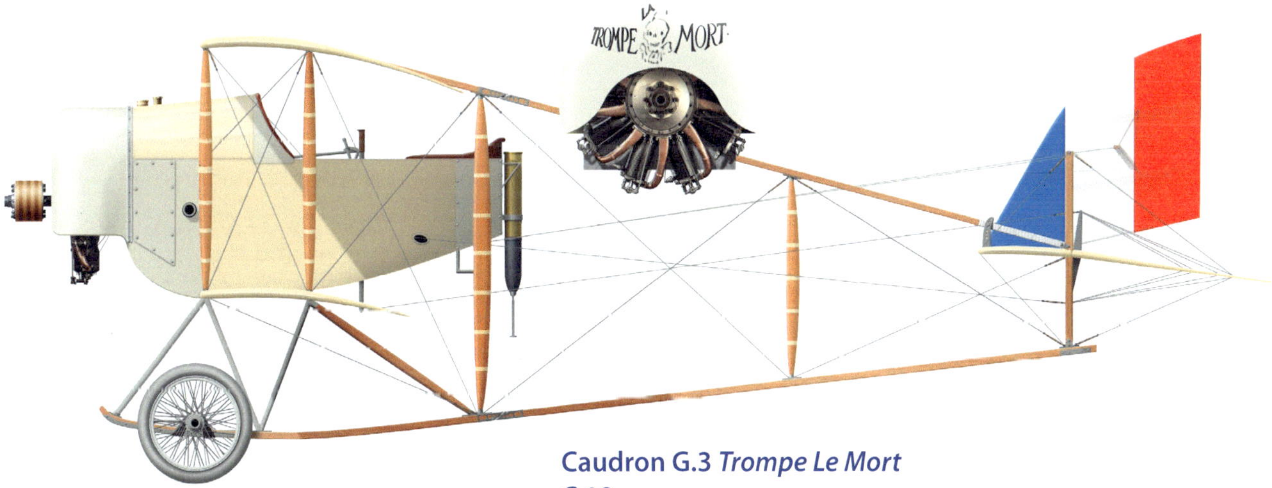

Caudron G.3 *Trompe Le Mort*
C 18
Autumn 1915

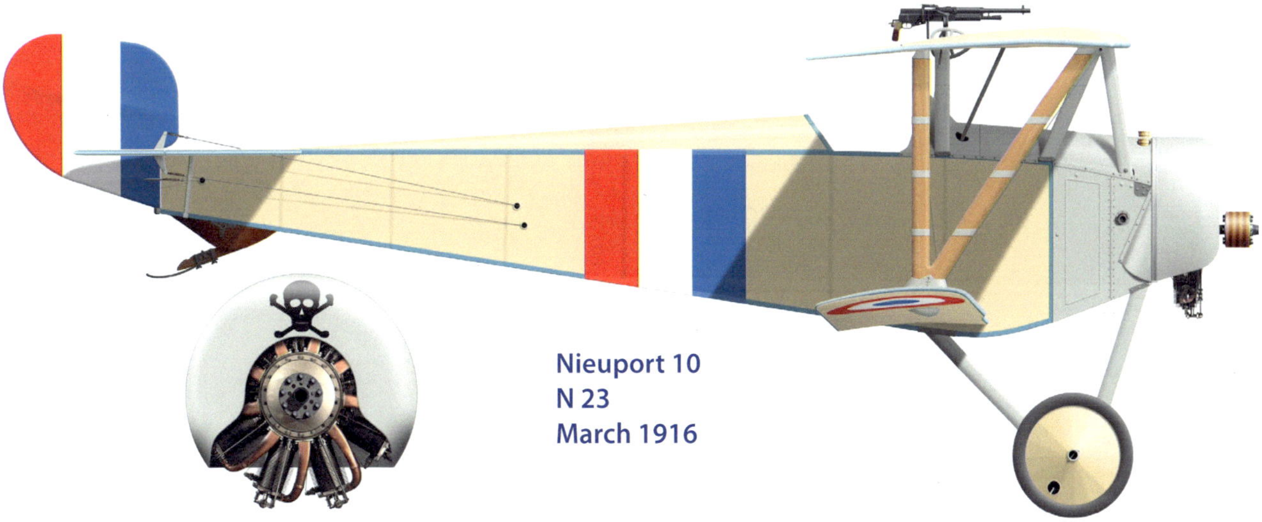

Nieuport 10
N 23
March 1916

Nieuport 16 *MAX* N977
N 23
May 1916

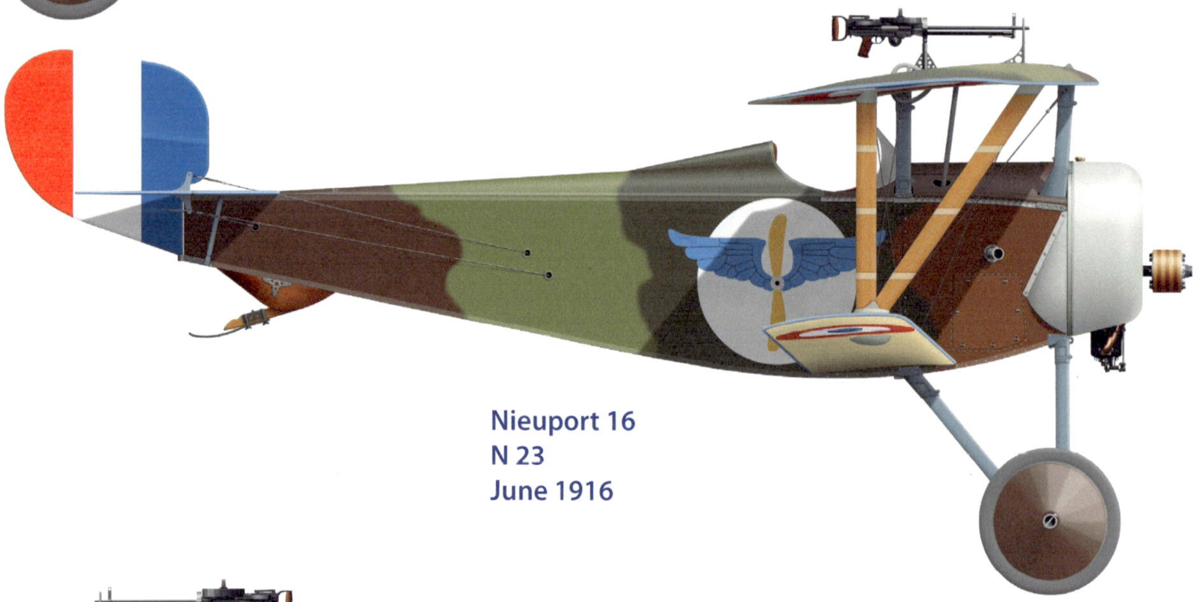

Nieuport 16
N 23
June 1916

Nieuport 16 *Alfred* N965
N 23
July 1916
Lenior occasionally flew this fighter.

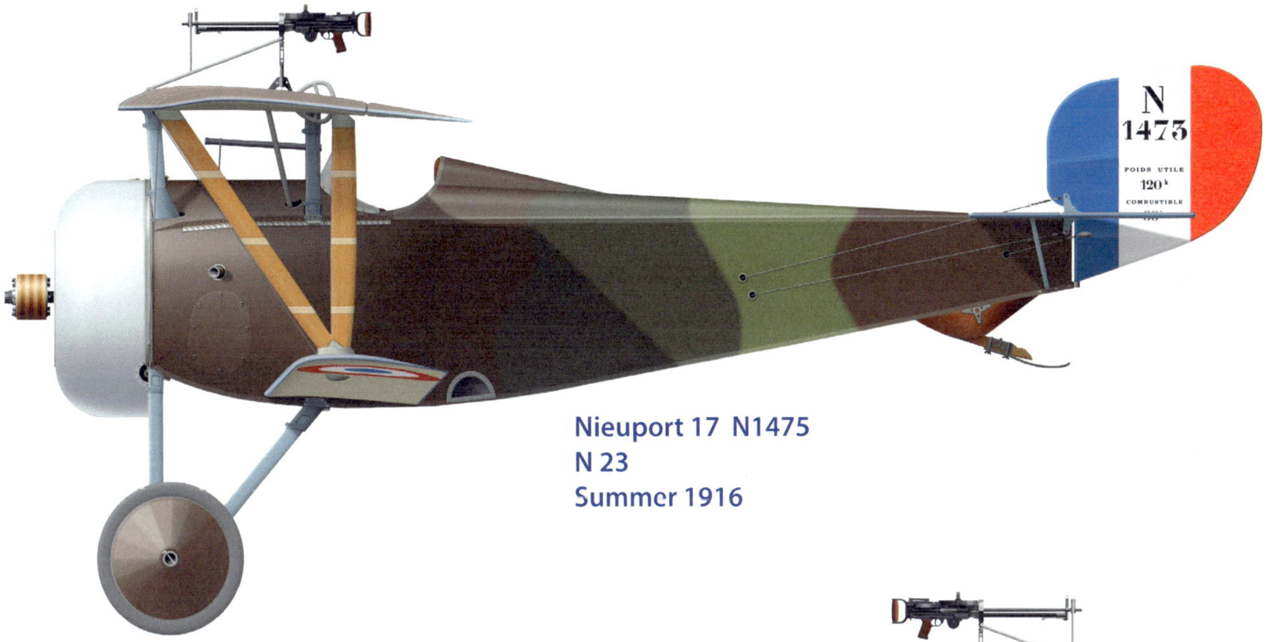

Nieuport 17 N1475
N 23
Summer 1916

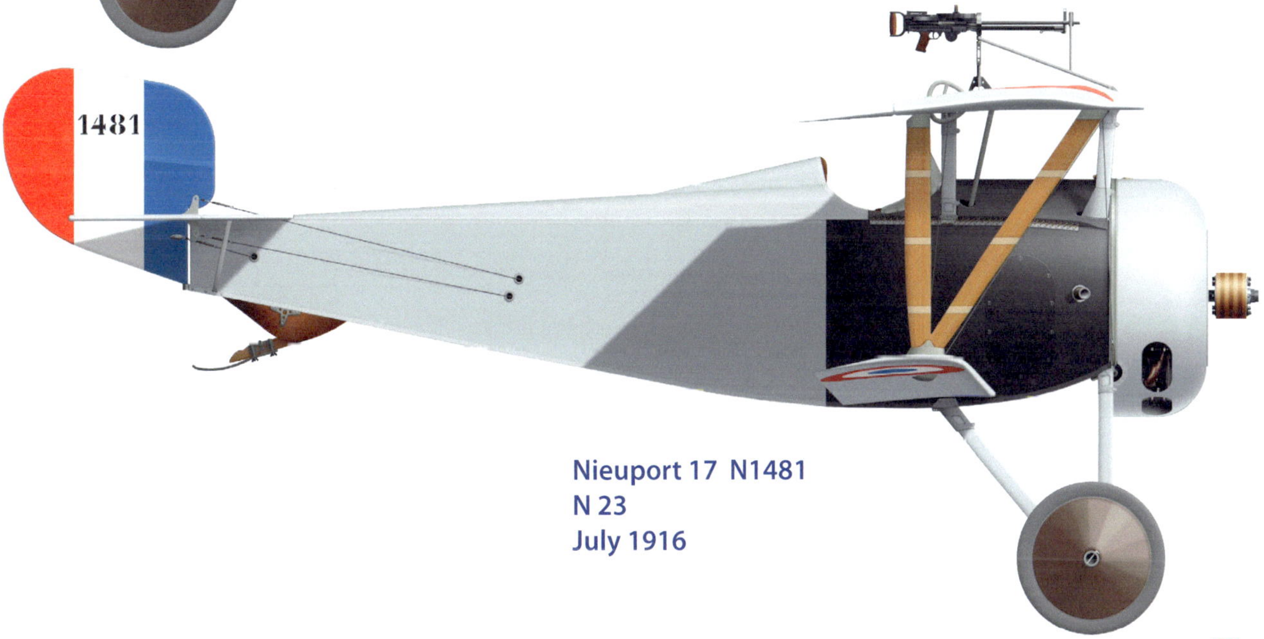

Nieuport 17 N1481
N 23
July 1916

Spad 7 *Trompe la Mort III* S116
N 23
October 1916

Adrien Louis Jacques Leps
(1893–1983)
Capitaine
N 67, Spa 81
12 victories

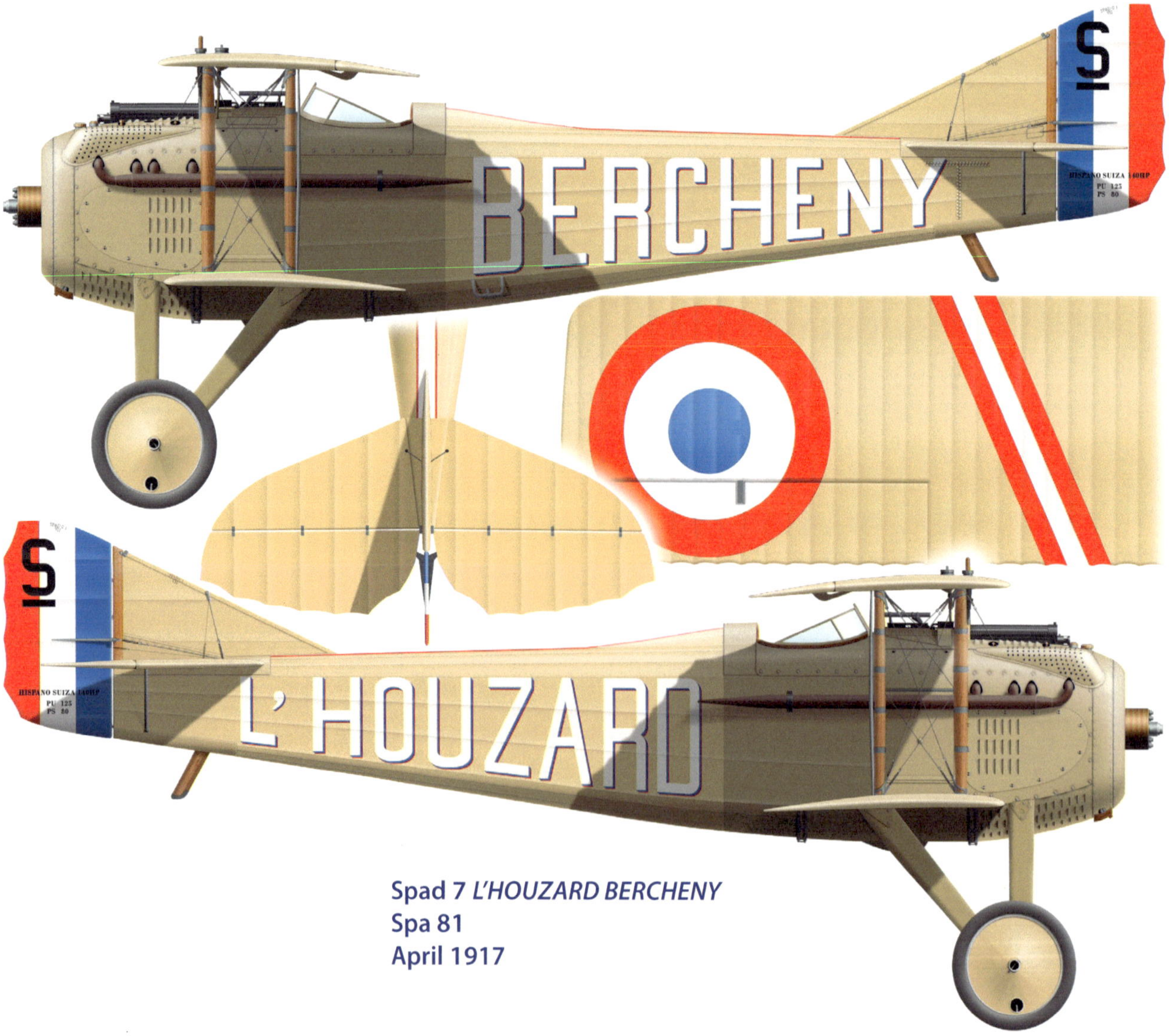

Spad 7 *L'HOUZARD BERCHENY*
Spa 81
April 1917

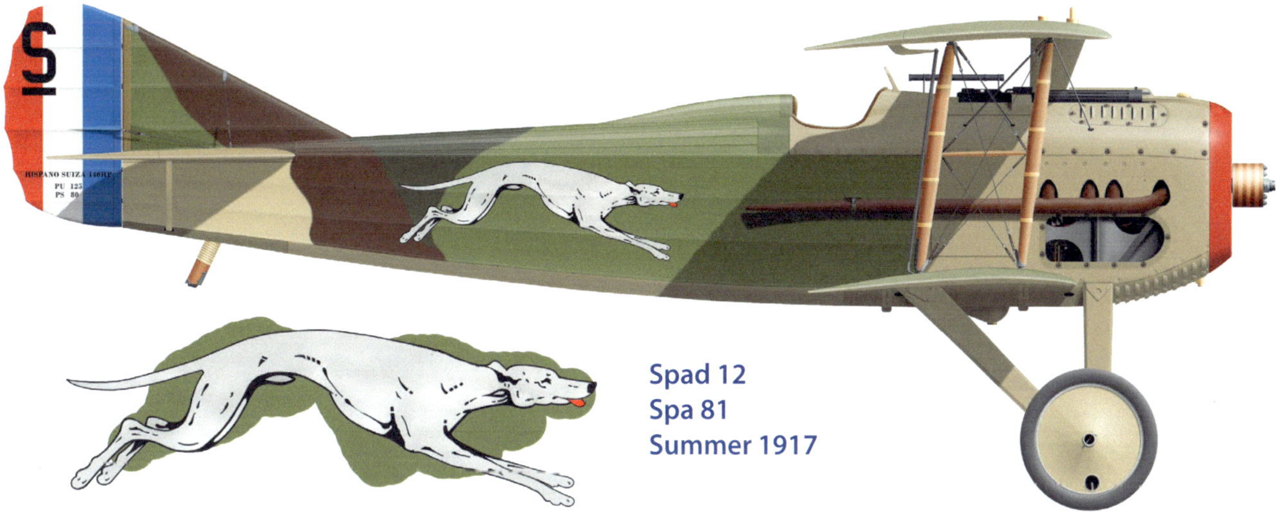

Spad 12
Spa 81
Summer 1917

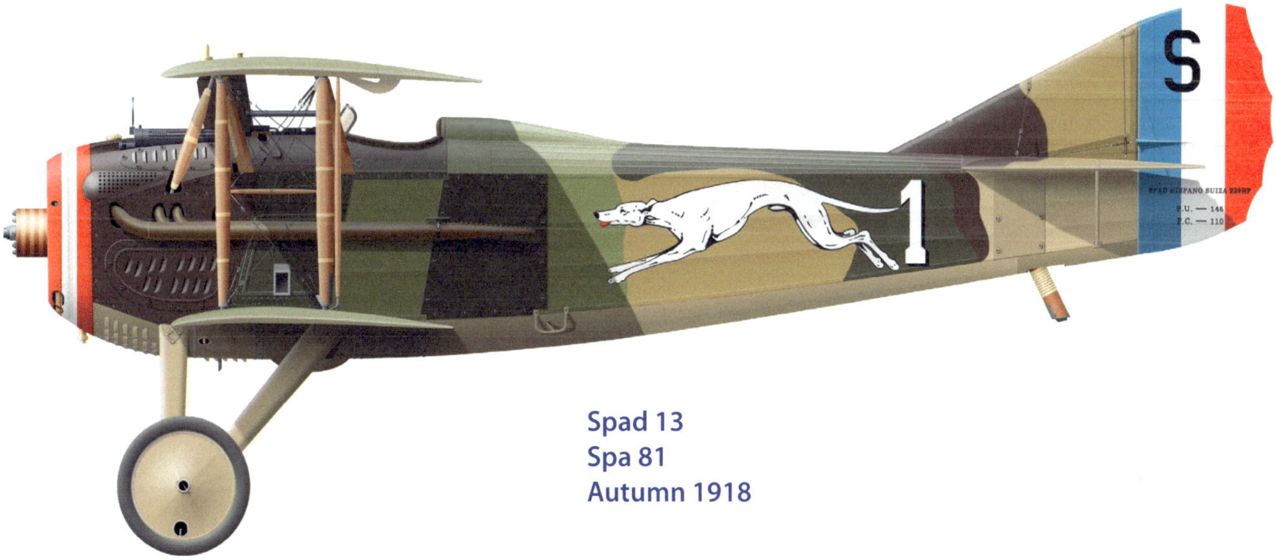

Spad 13
Spa 81
Autumn 1918

André Robert Lévy
(1893-1973)
Sergent
Spa 561
6 victories

Nieuport 17 N2756
N 561
Spring 1917

Nieuport 17 *Chiff!* N4663
N 561
Autumn 1917
Sous-Lt. Jean Bignon also flew this airplane.

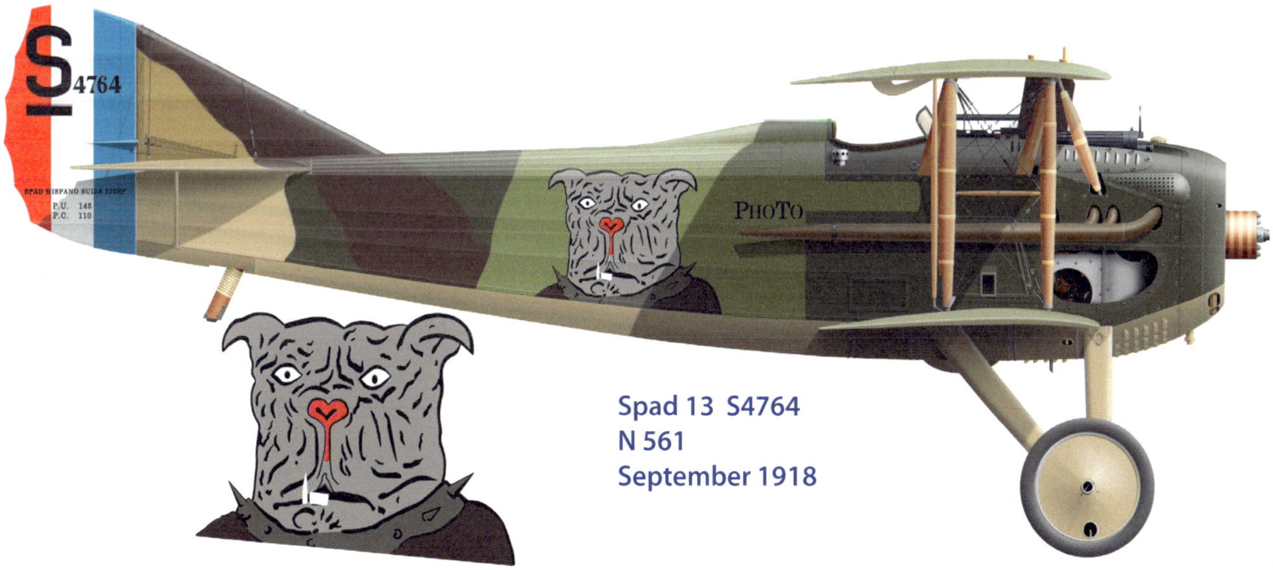

Spad 13 S4764
N 561
September 1918

Georges Felix Madon
(1892–1924)
Capitaine
BL 30, MF 218, N 38, Spa 38
41 victories

Bleriot 11 BL231
BL 30
Autumn 1914

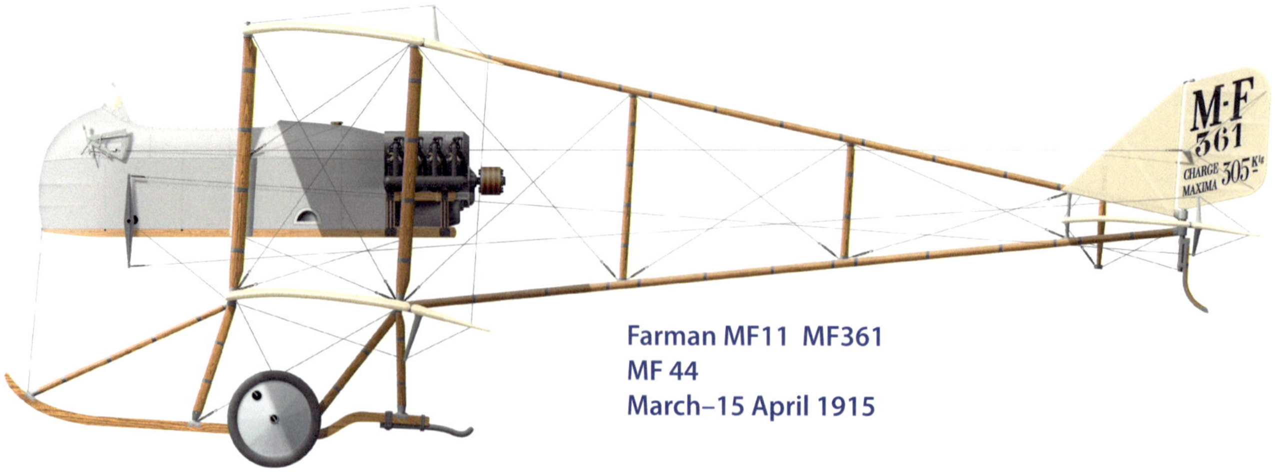

Farman MF11 MF361
MF 44
March–15 April 1915

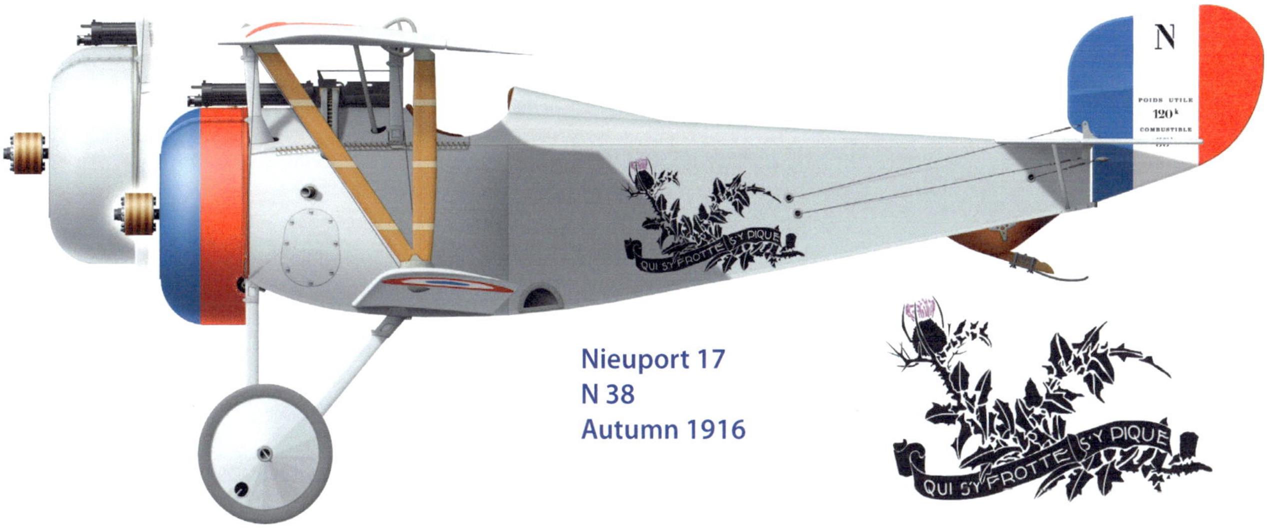

Nieuport 17
N 38
Autumn 1916

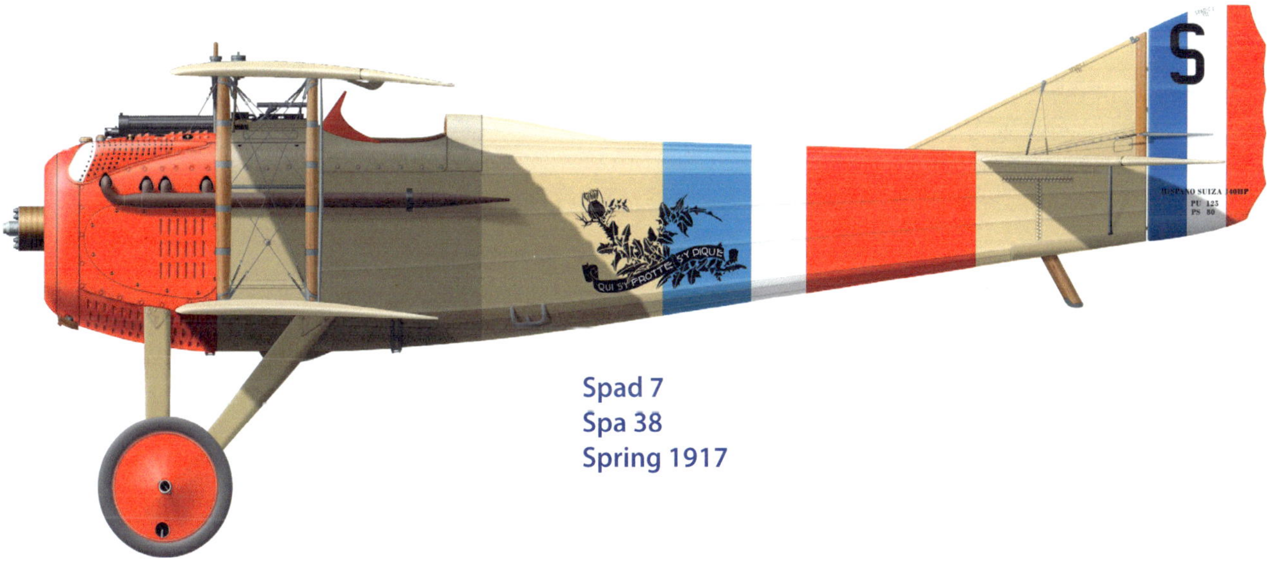

Spad 7
Spa 38
Spring 1917

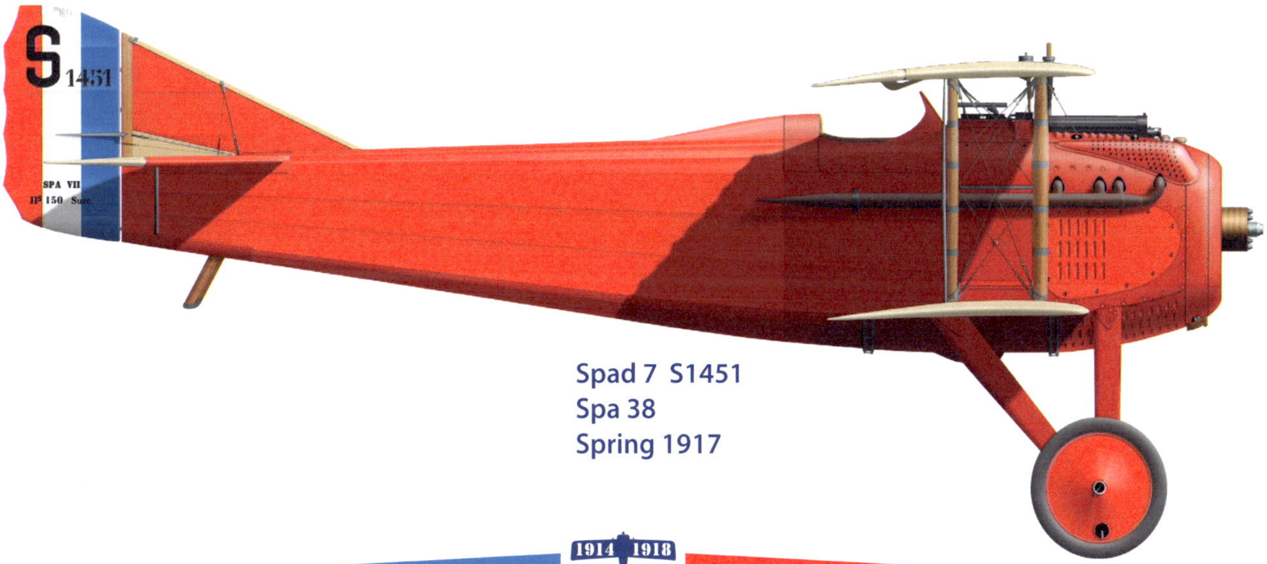

Spad 7 S1451
Spa 38
Spring 1917

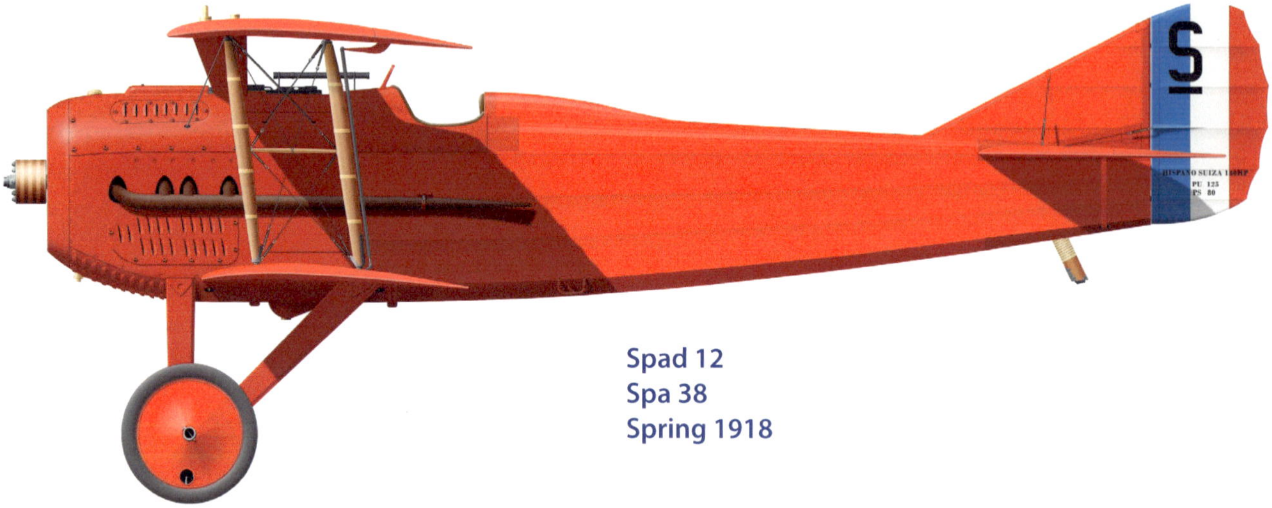

Spad 12
Spa 38
Spring 1918

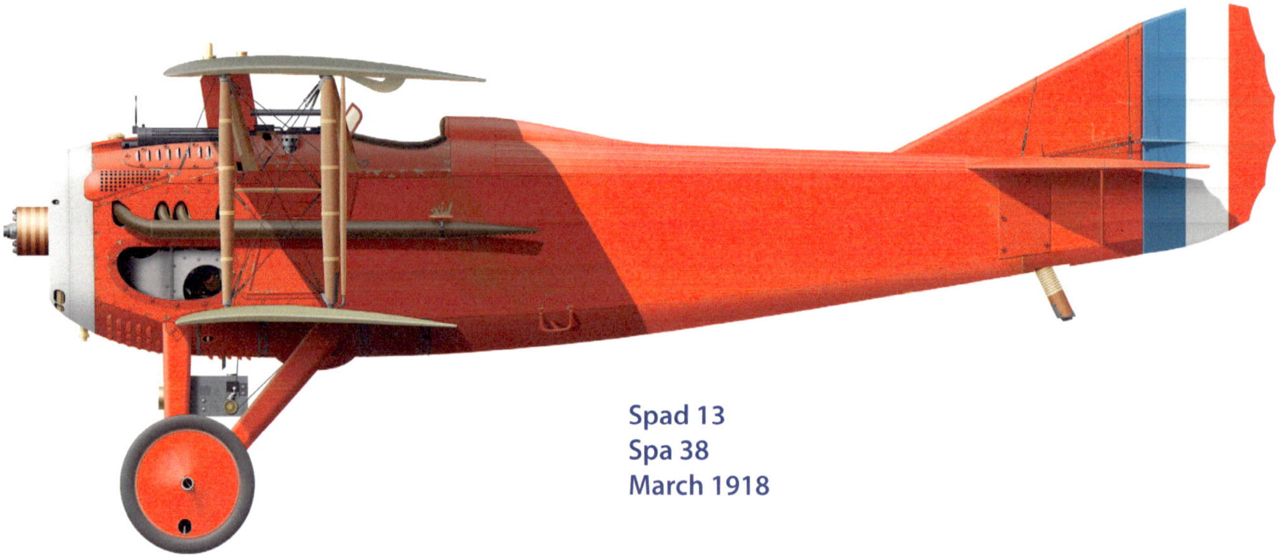

Spad 13
Spa 38
March 1918

Nieuport Monoplane
Spa 38
July 1918

Paul Alphonse Louis Napoléon Malavialle (1888–1919)

Capitaine
BLC 5, N 57, N 69
5 victories

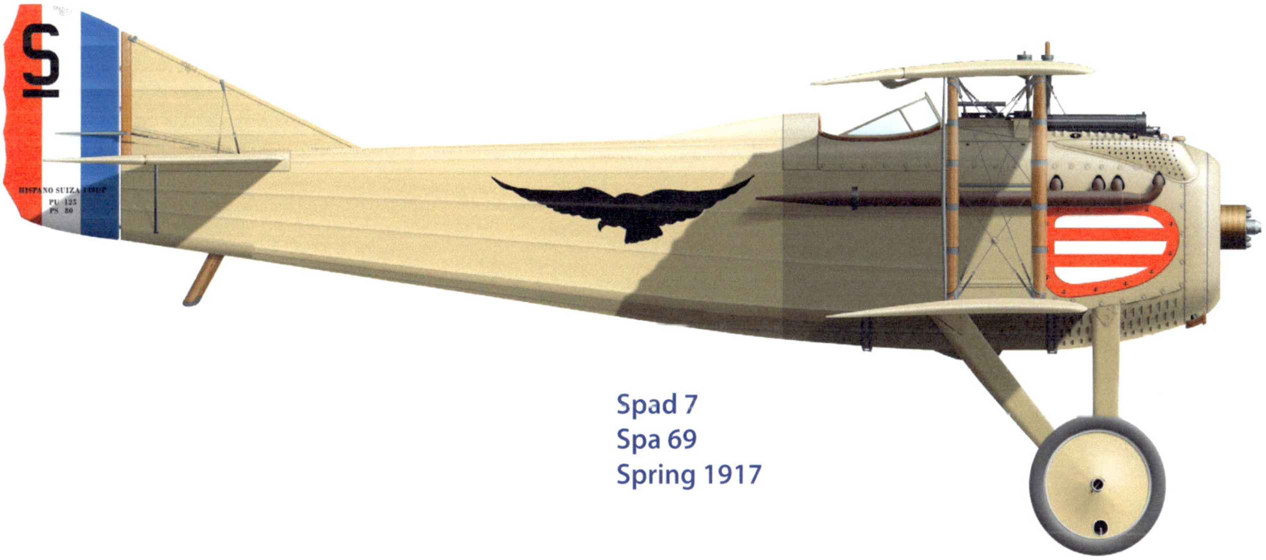

Spad 7
Spa 69
Spring 1917

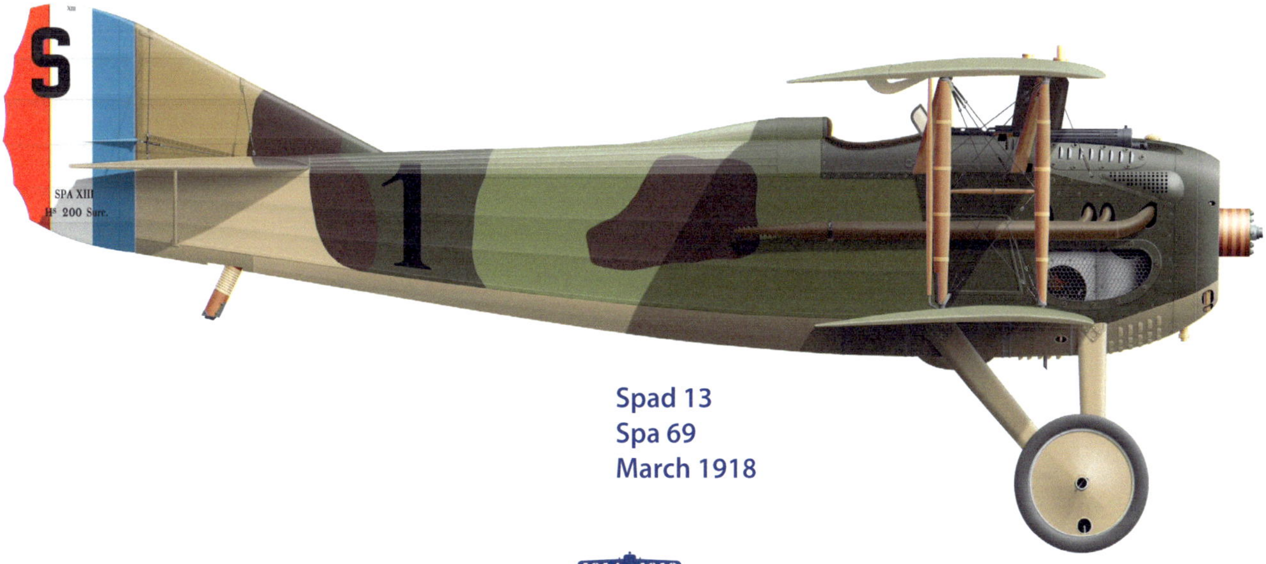

Spad 13
Spa 69
March 1918

Pierre Marinovitch

(1898–1919)
Sous-Lieutenant
N 38, Spa 94
21 victories

Nieuport 24*bis* N3317
N 94
September 1917

Spad 13
Spa 94
Spring 1918

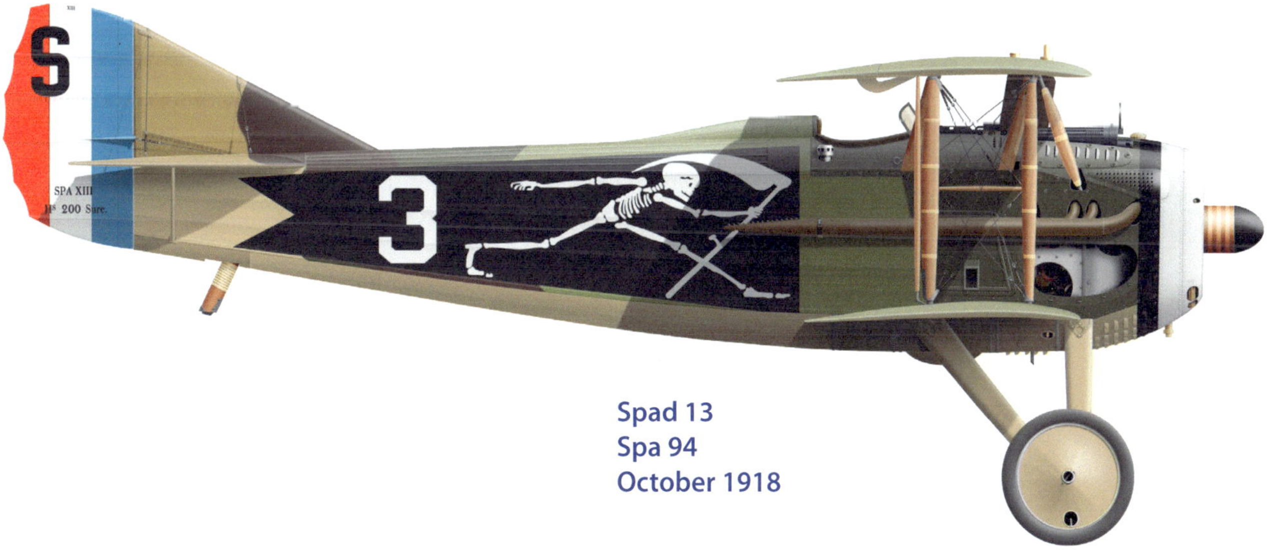

Spad 13
Spa 94
October 1918

Lionel Alexandre Pierre de Marmiere
(1897–1944)
Sous-Lieutenant
Spa 112, Spa 176
6 victories

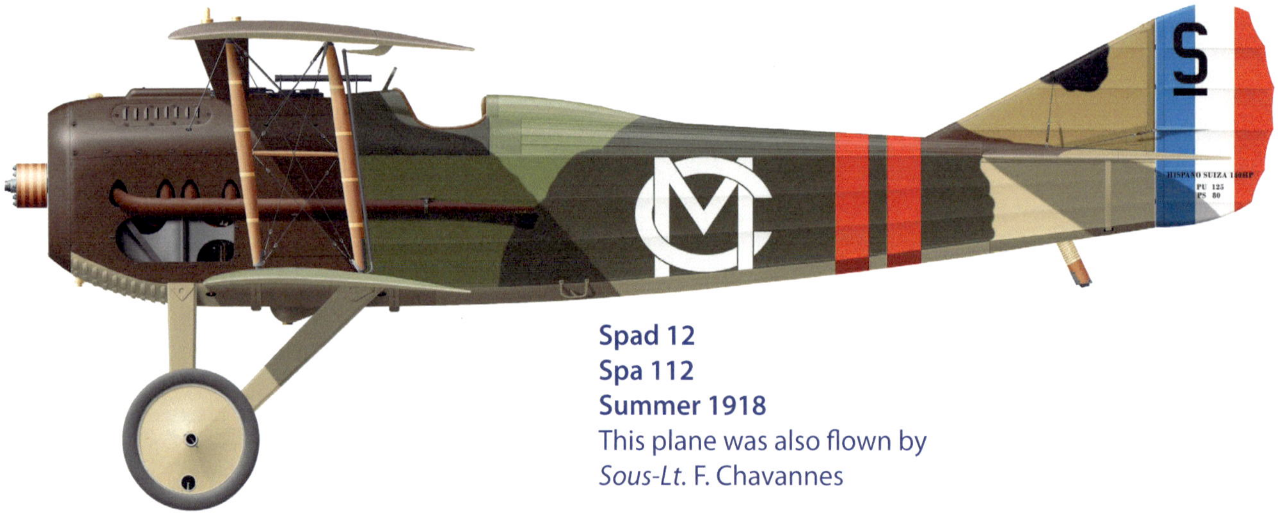

Spad 12
Spa 112
Summer 1918
This plane was also flown by
Sous-Lt. F. Chavannes

Caudron-Renault 714 No. 57 (I-247)
GC I/145 (*Grope Chasse* I "*Warszawa*")
June 1940

Andre Henri Martenot de Cordou
(1893–1991)
Sous-Lieutenant
C 28, C 56, N 513, N 38, N 94, Spa 94
7 or 8 victories

Nieuport 16
N 38
May 1916

Nieuport 24
N 94
July 1917

Spad 7 S135
N 94
September 1917

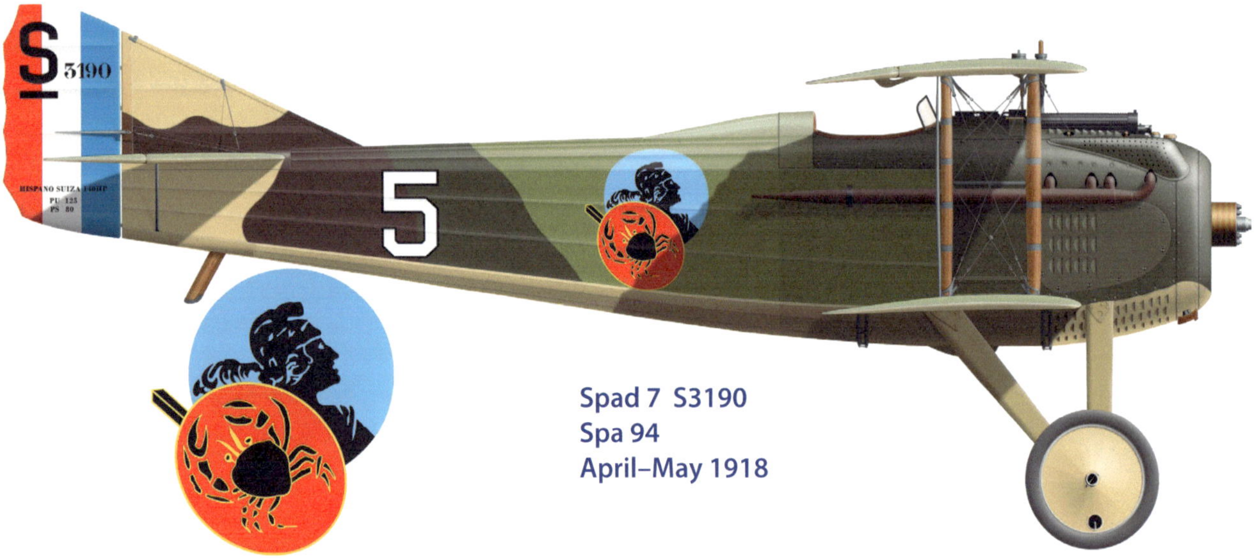

Spad 7 S3190
Spa 94
April–May 1918

Spad 7 S3190?
Spa 94
Summer 1918

Alexandre Paul Léon Madeleine Marty
(1894–1918)
Lieutenant
Spa 77, Spa 90
7 victories

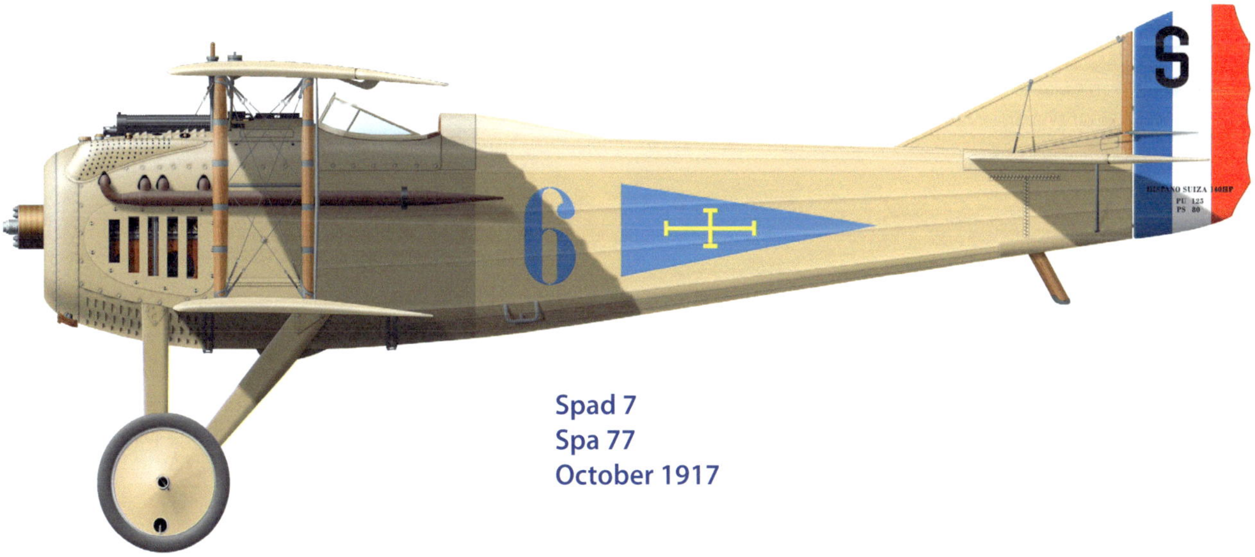

Spad 7
Spa 77
October 1917

Robert Léon Henri Massenet-Royer de Marancour (1880–1969)

Chef-de-Bataillon

BLC 5, N 69, GC 14

8 victories

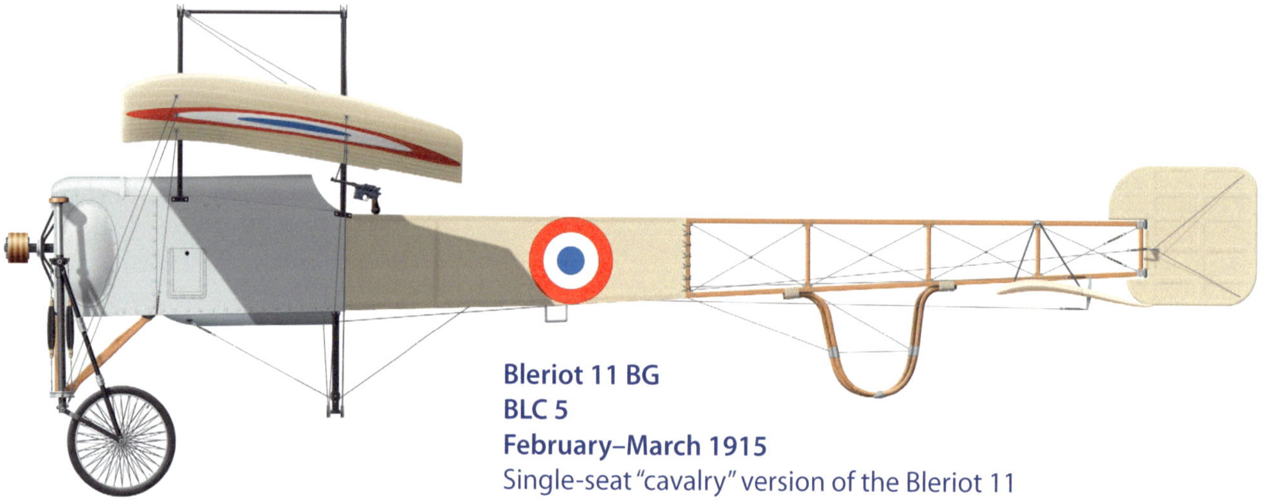

Bleriot 11 BG
BLC 5
February–March 1915
Single-seat "cavalry" version of the Bleriot 11

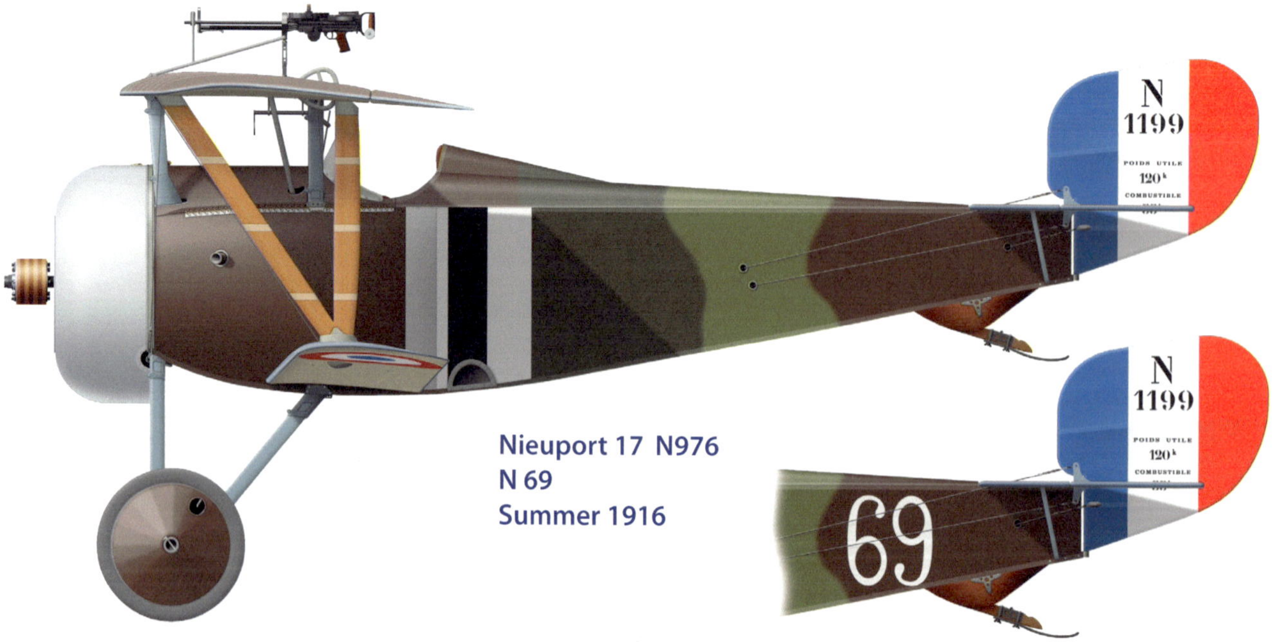

Nieuport 17 N976
N 69
Summer 1916

Jean Georges Fernand Matton
(1888–1917)
Capitaine
MF 20, N 48, N 57, N 77
9 victories

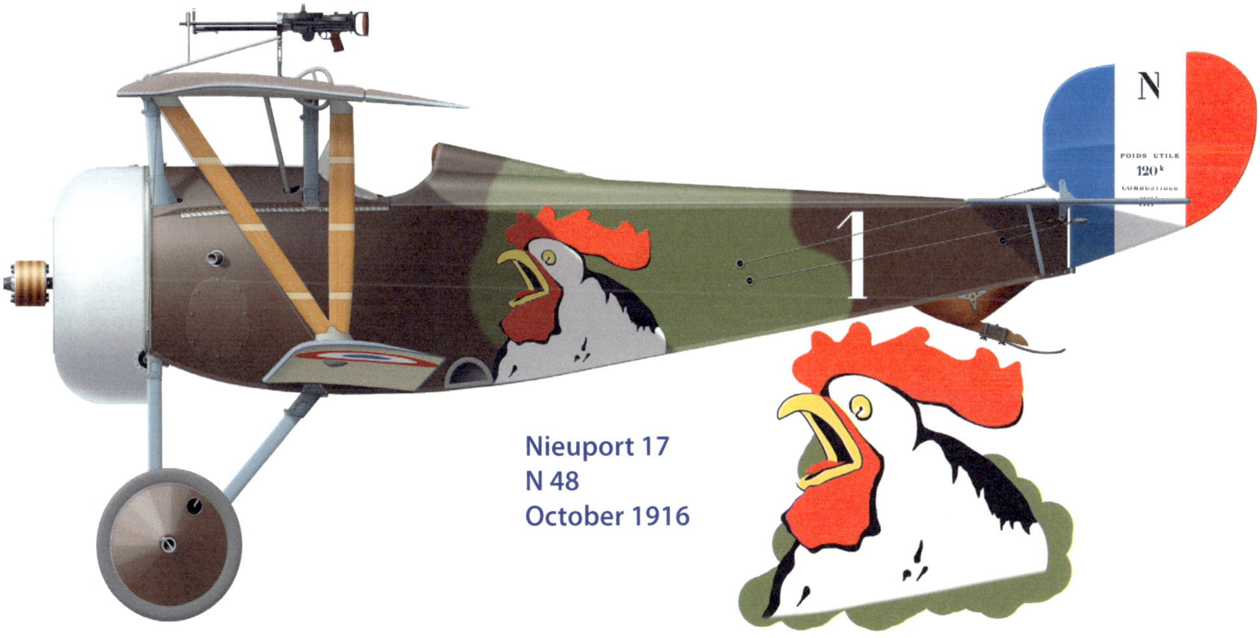

Nieuport 17
N 48
October 1916

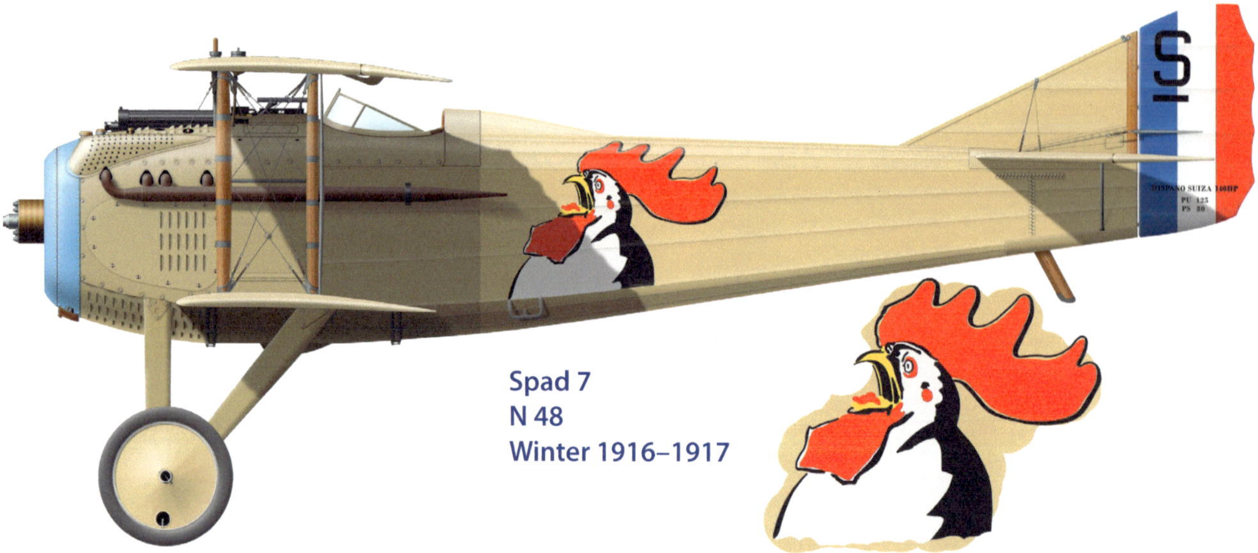

Spad 7
N 48
Winter 1916–1917

Spad 7
N 48
Summer 1917

Ernest Joseph Jules Maunoury
(1894–1921)
Sous-Lieutenant
C 64, F 16, N 152, Spa 152
11 victories

Nieuport 24*bis*
Spa 152
Autumn 1917

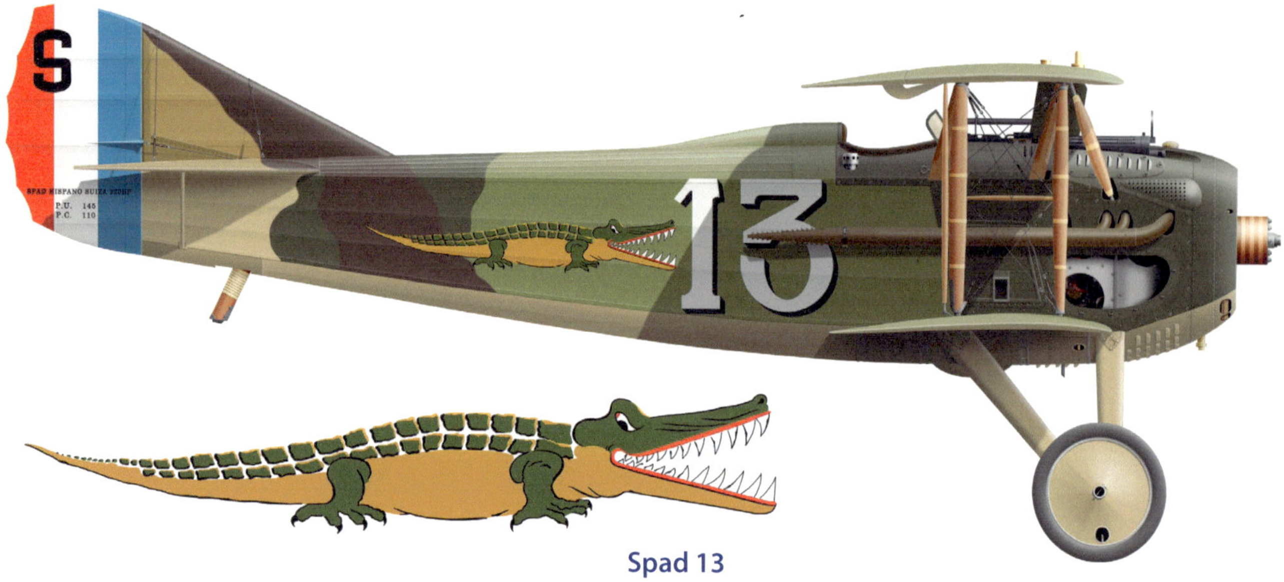

Spad 13
Spa 152
Summer 1918

Spad 13 *Simon* S77??
Spa 152
Autumn 1918

Albert Edmond Mezergues
(1886–1925)
Lieutenant
V 21, VC 113, V 90S, V 84S, BR 131, Esc 21, N 79, Sop 129
6 victories

Voisin 4 V985
V 90S
May 1916

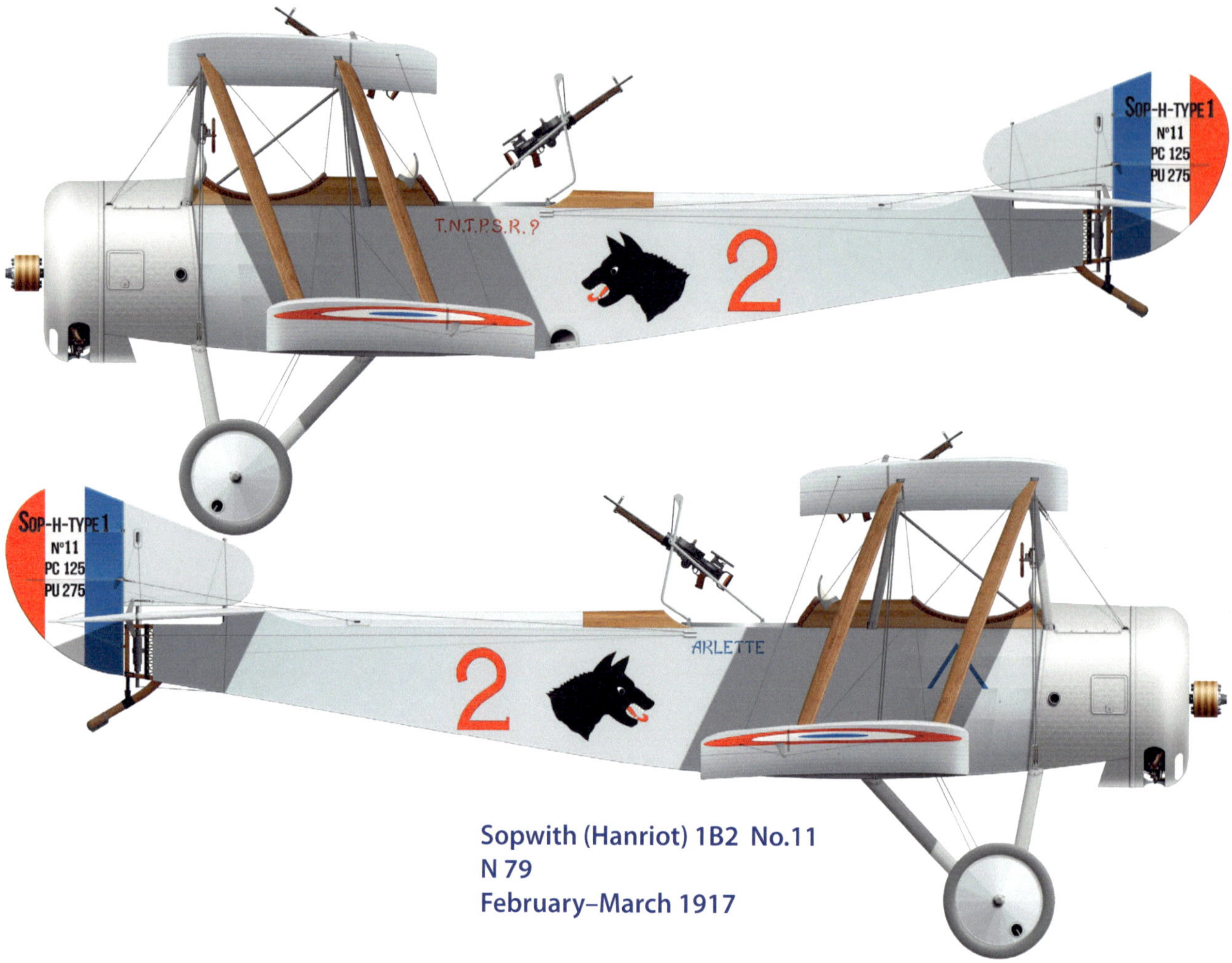

Sopwith (Hanriot) 1B2 No.11
N 79
February–March 1917

Sopwith 1B1 No.145
N 79
June 1917

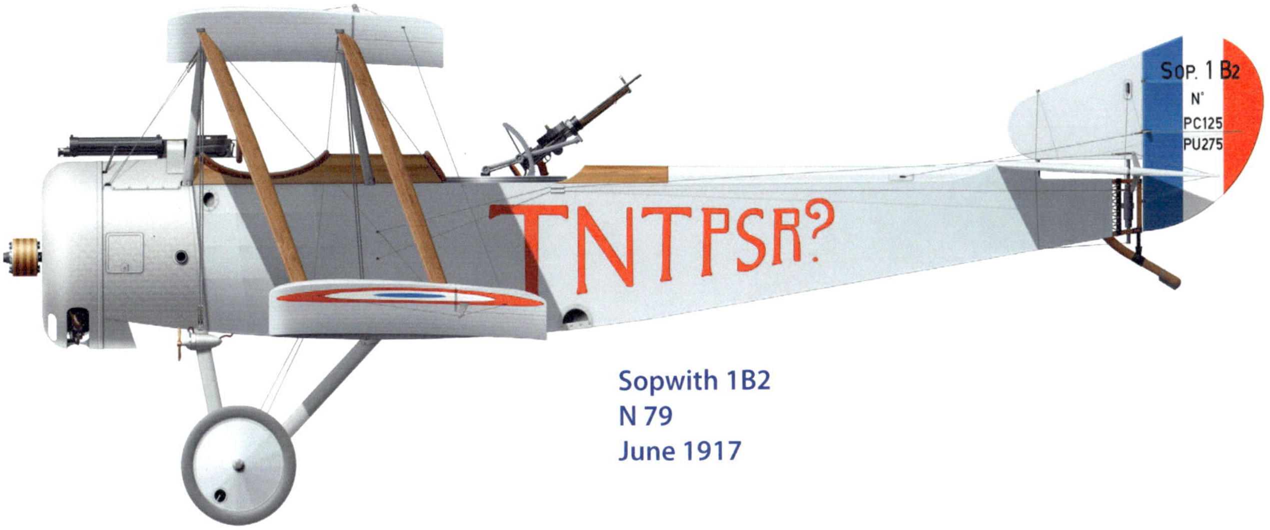

Sopwith 1B2
N 79
June 1917

Sopwith 1B2 *Arlette*
N 79
May–June 1917
This aircraft was not assigned to Mezergues, but he occasionally flew it.

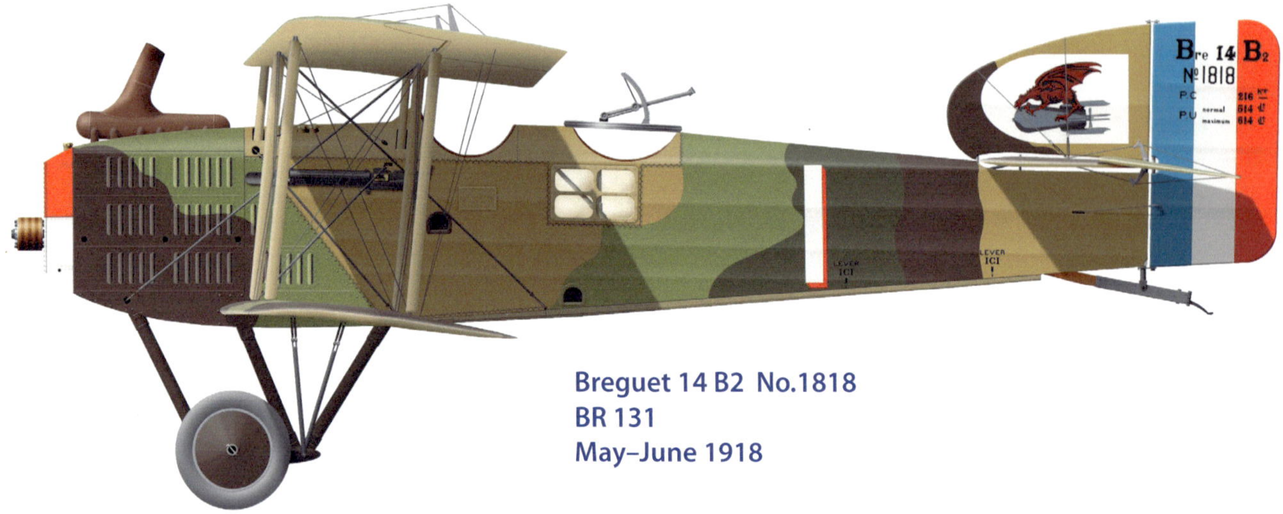

Breguet 14 B2 No.1818
BR 131
May–June 1918

Breguet 14 B2 No.4070
BR 131
June 1918

Xavier Jean-Marie Louis Moissinac
(1896–1918)
Marechal-des-Logis
Spa 154
7 victories

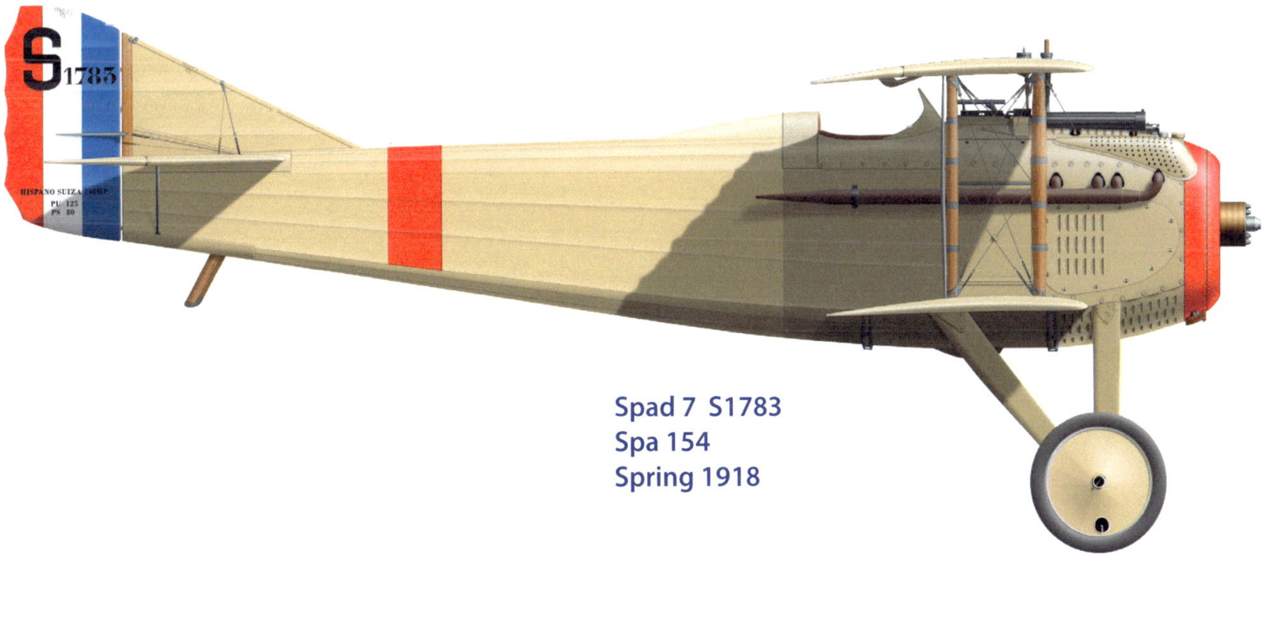

Spad 7 S1783
Spa 154
Spring 1918

Paul Montange
(1892–1963)
Sergent
N 155, Spa 155
5 victories

Nieuport 24*bis Suzy* N4495
N 155
September 1917

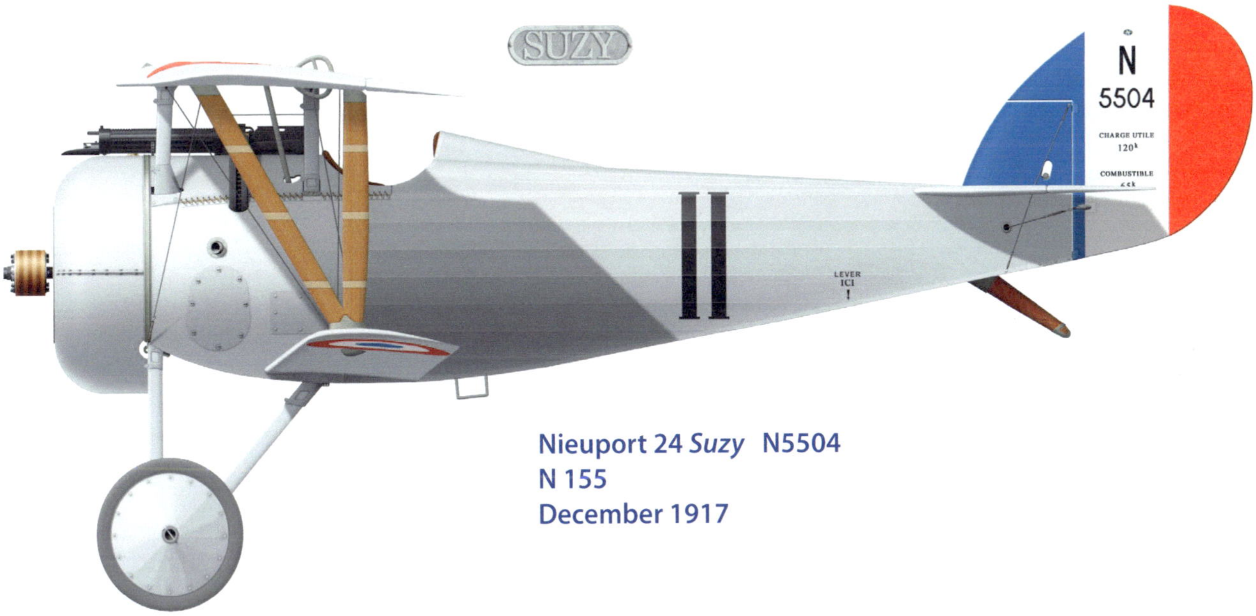

Nieuport 24 *Suzy* N5504
N 155
December 1917

Spad 13 *Suzy* S3837
Spa 154
September 1918

René Montrion
(1896–1918)
Adjudant
Spa 48
11 victories

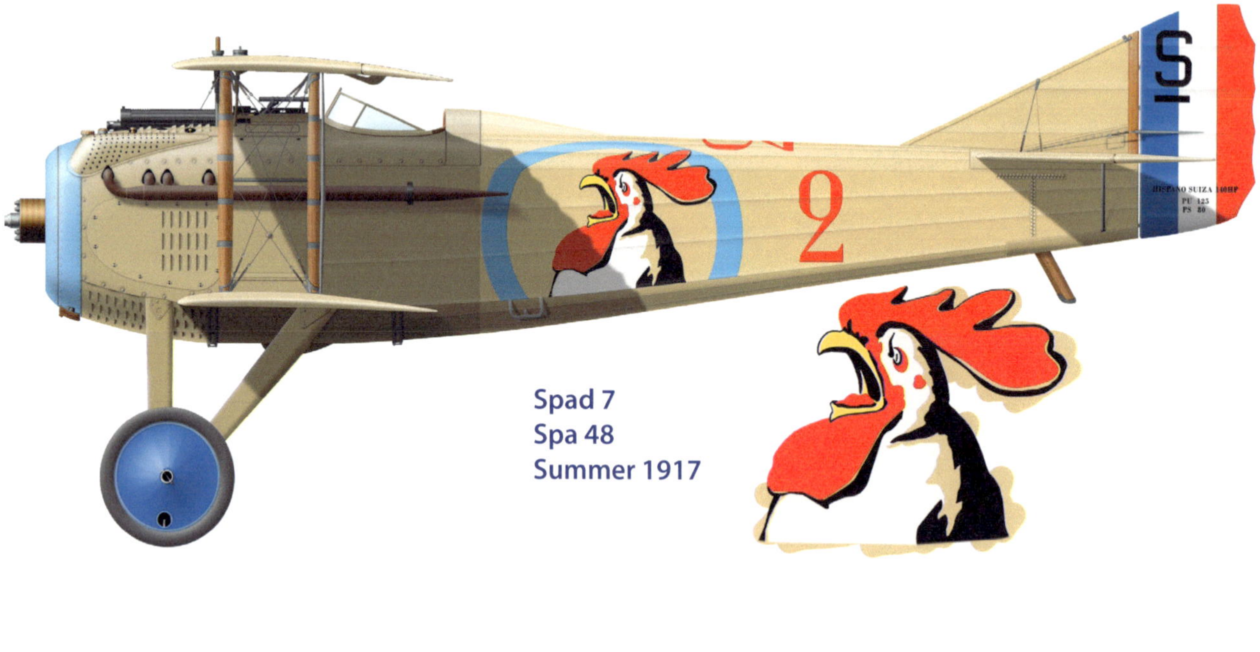

Spad 7
Spa 48
Summer 1917

Jean Marie Dominique Navarre
(1895–1919)
Sous-Lieutenant
MF 8, N 12, MS 12, N 67
12 victories

Morane Saulnier L MS27
MS 12
Spring 1915

Morane-Saulnier LA *Rouli-Rouli*
MS 12
Summer 1915

Morane-Saulnier N MS398
MS 12
Autumn 1915

Morane-Saulnier N
MS 12
Autumn 1915

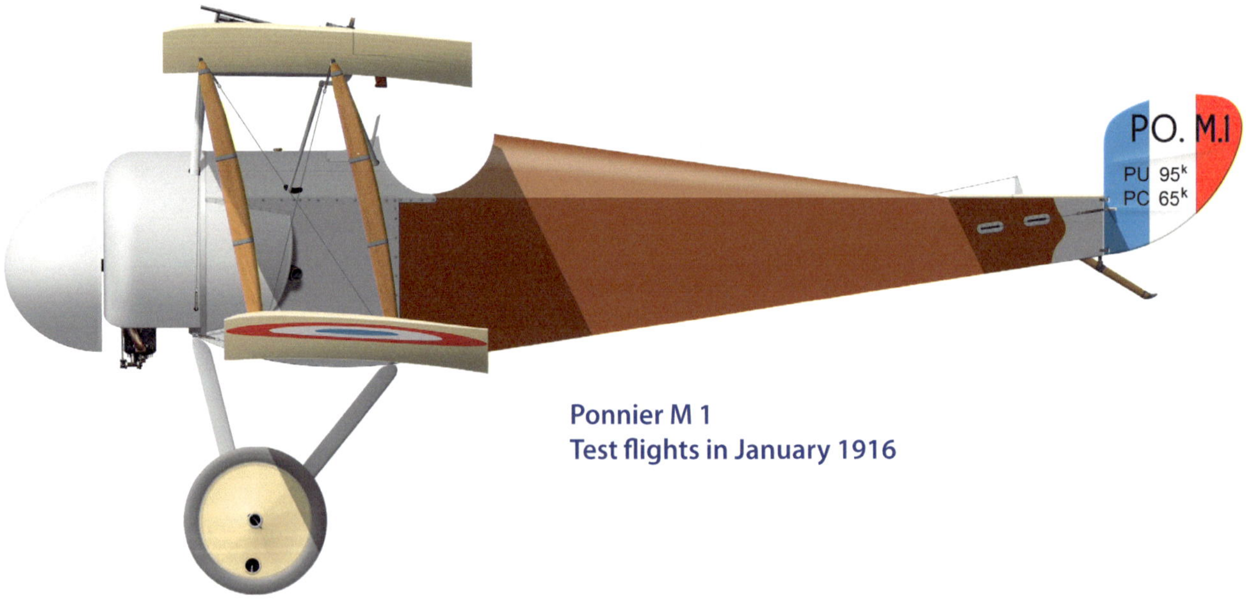

Ponnier M 1
Test flights in January 1916

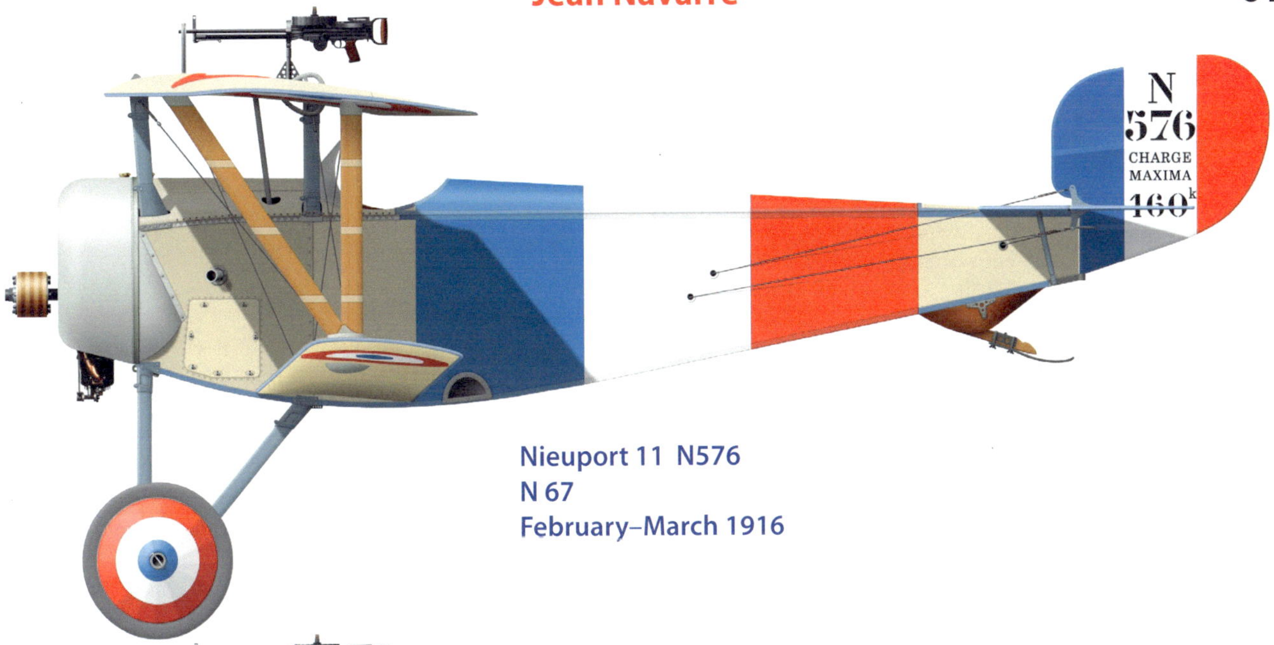

Nieuport 11 N576
N 67
February–March 1916

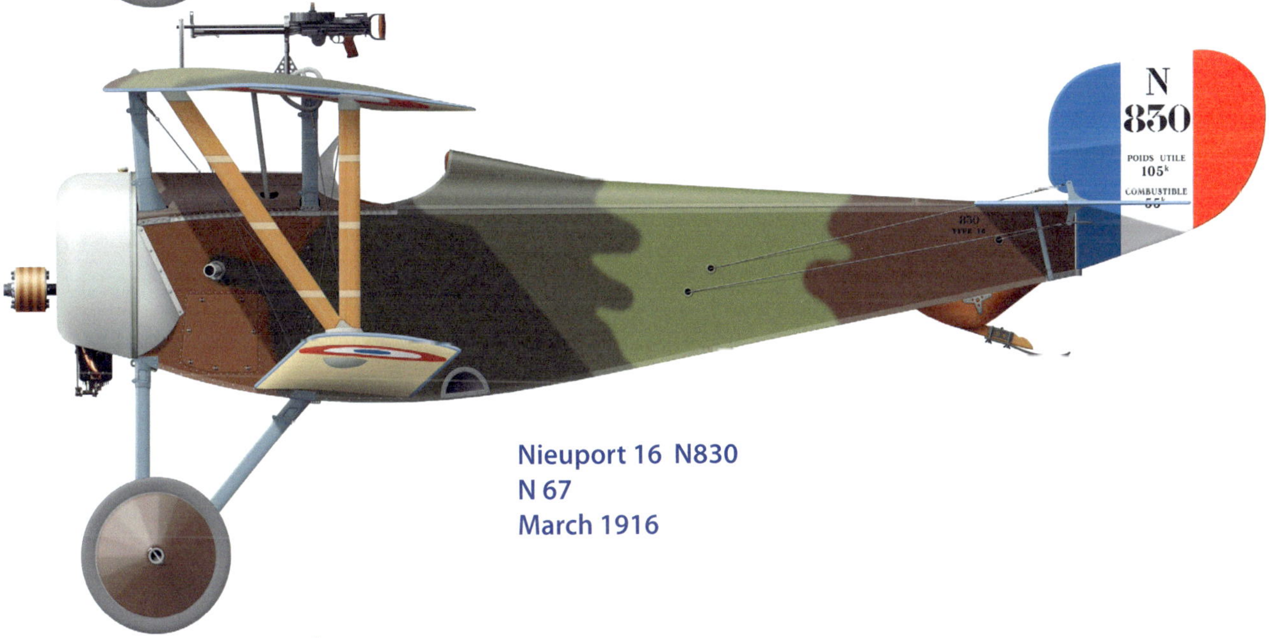

Nieuport 16 N830
N 67
March 1916

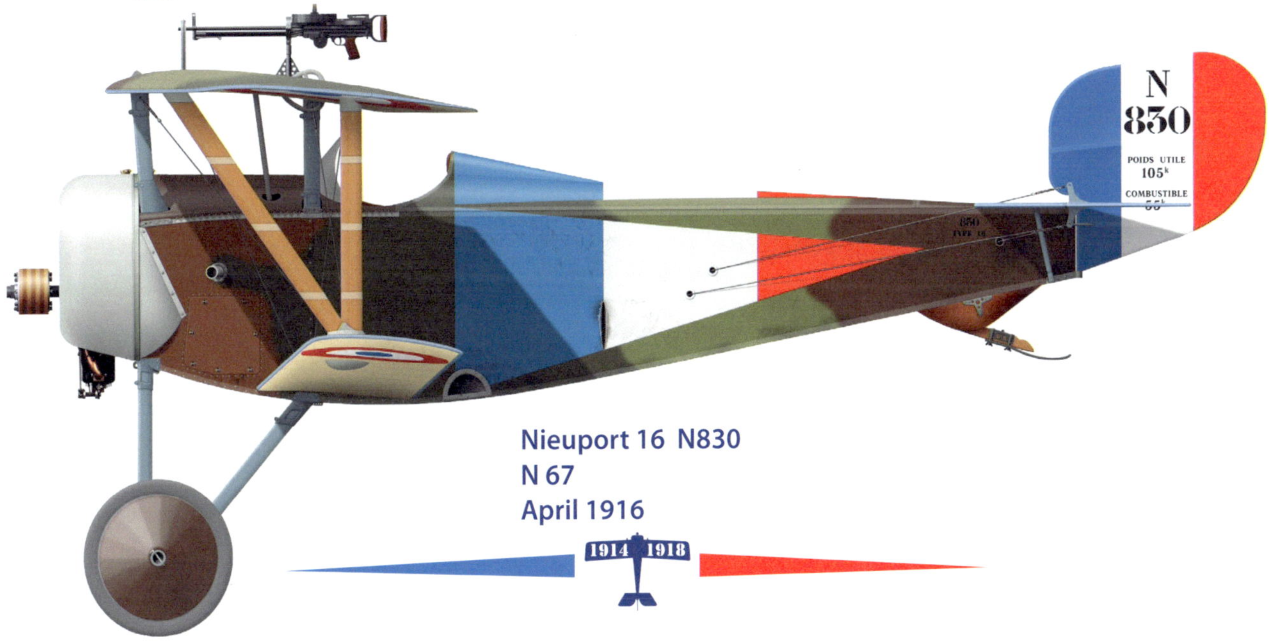

Nieuport 16 N830
N 67
April 1916

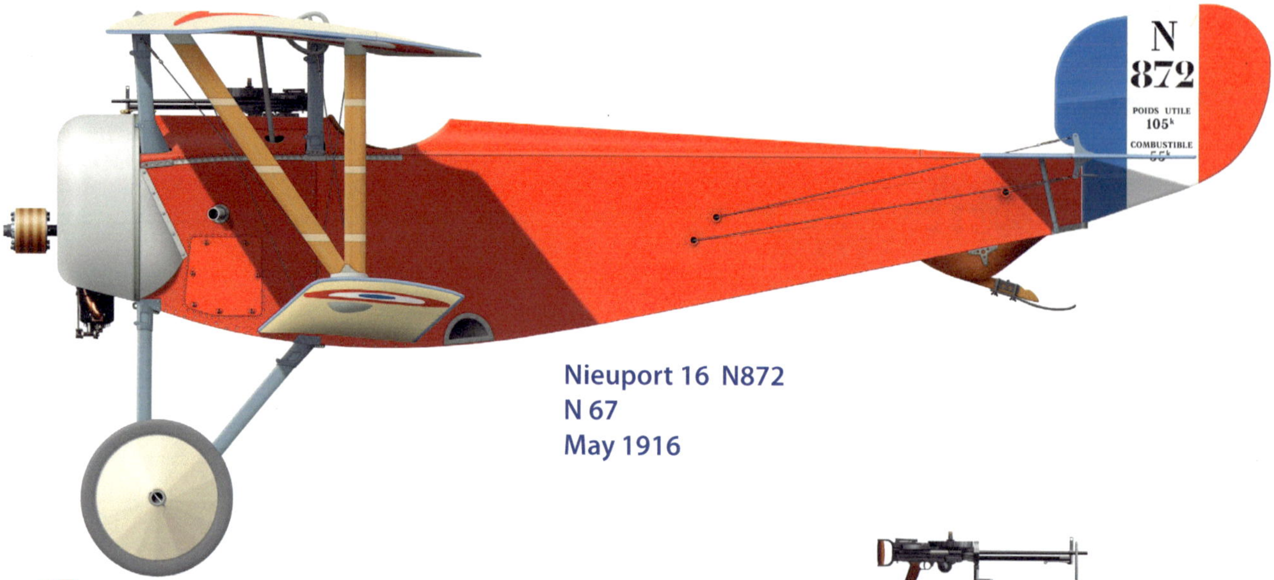

Nieuport 16 N872
N 67
May 1916

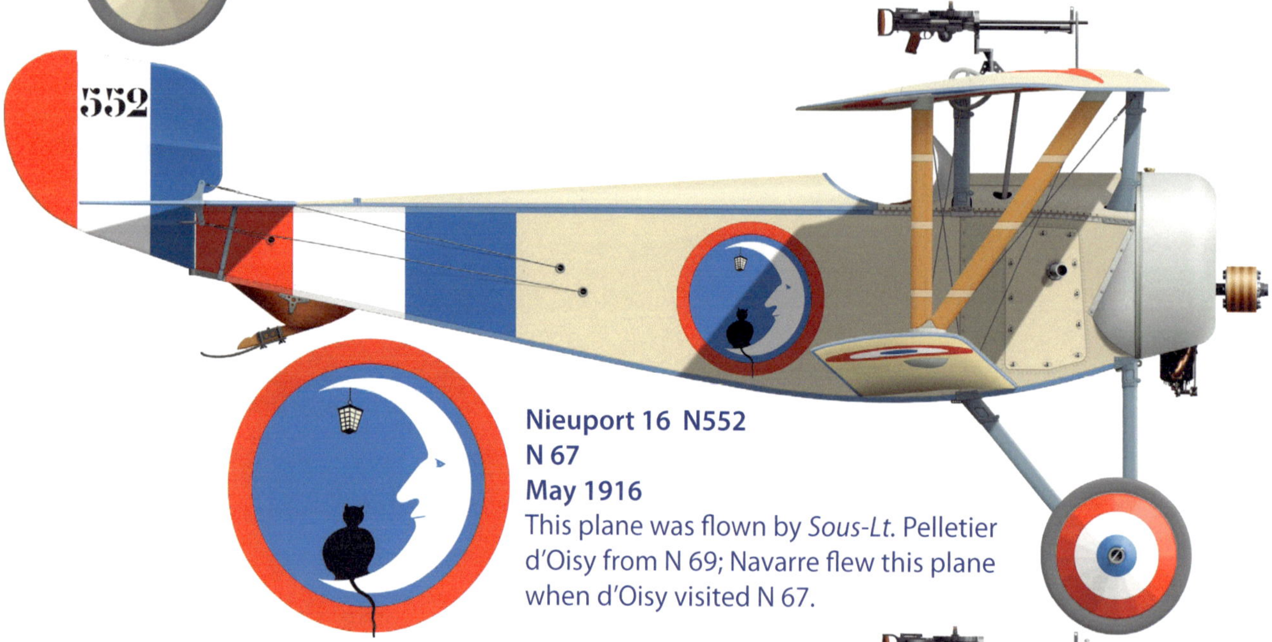

Nieuport 16 N552
N 67
May 1916
This plane was flown by *Sous-Lt.* Pelletier d'Oisy from N 69; Navarre flew this plane when d'Oisy visited N 67.

Nieuport 16
H•SS Allons!…tant mieux
N 67
May 1916

Nieuport 16 *H•SS Allons!…tant mieux*
N 67
May 1916
This fighter was flown by *Capt*. Henri de Saint-Sauveur, commander of N 67, but Navarre occasionally flew it.

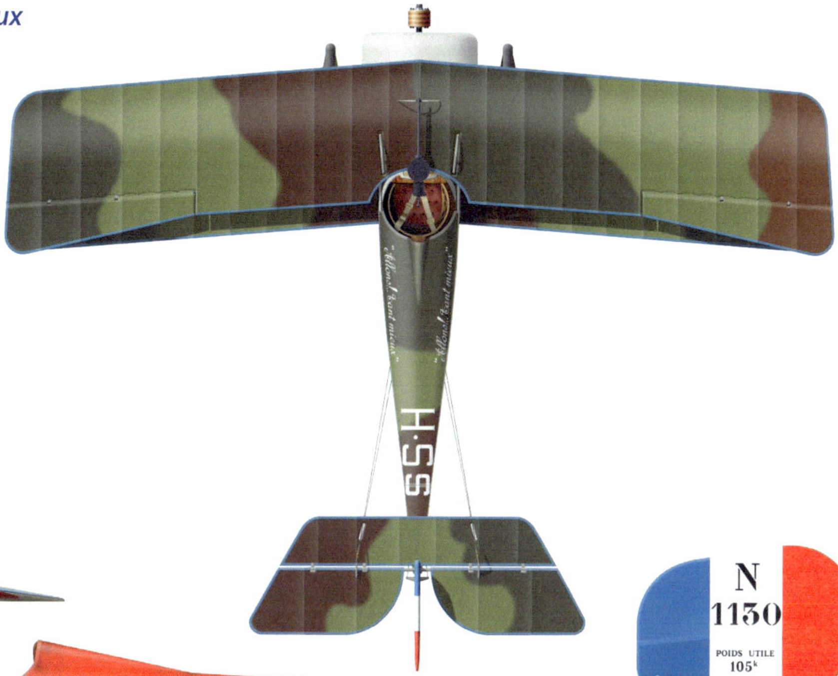

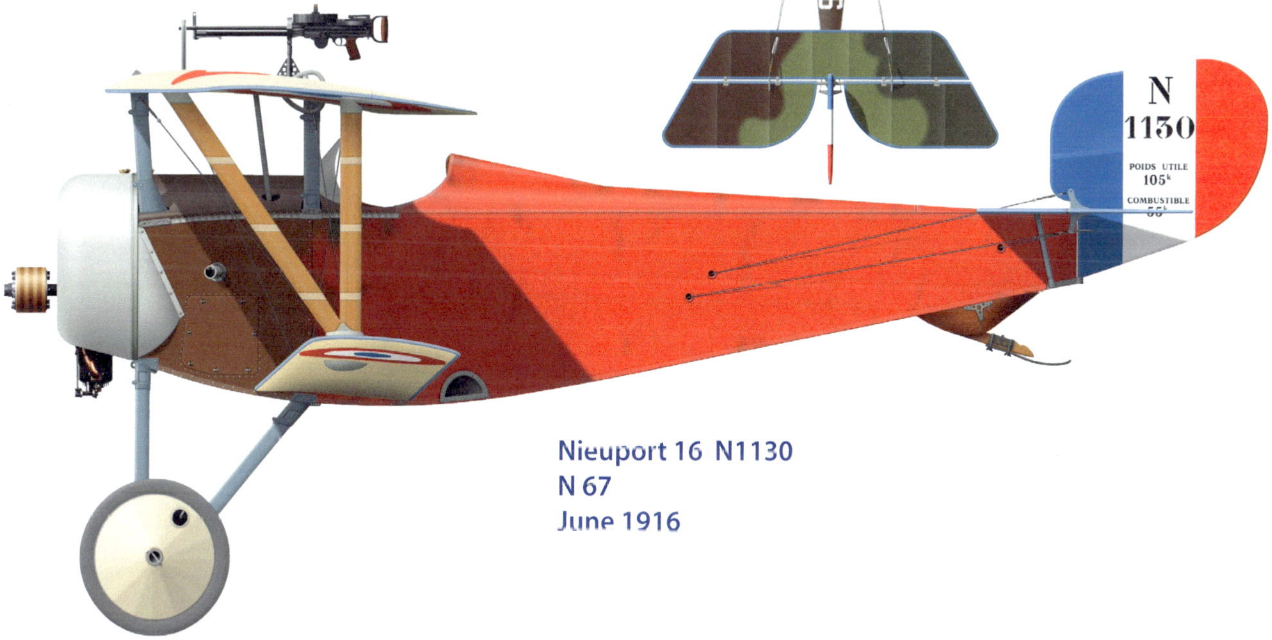

Nieuport 16 N1130
N 67
June 1916

Morane-Saulnier AI MS2083
Villacoublay Airfield
July 1919

Marcel Joseph Maurice Nogues
(1895–1919)
Sous-Lieutenant
N12, Spa57, Spa172
13 victories

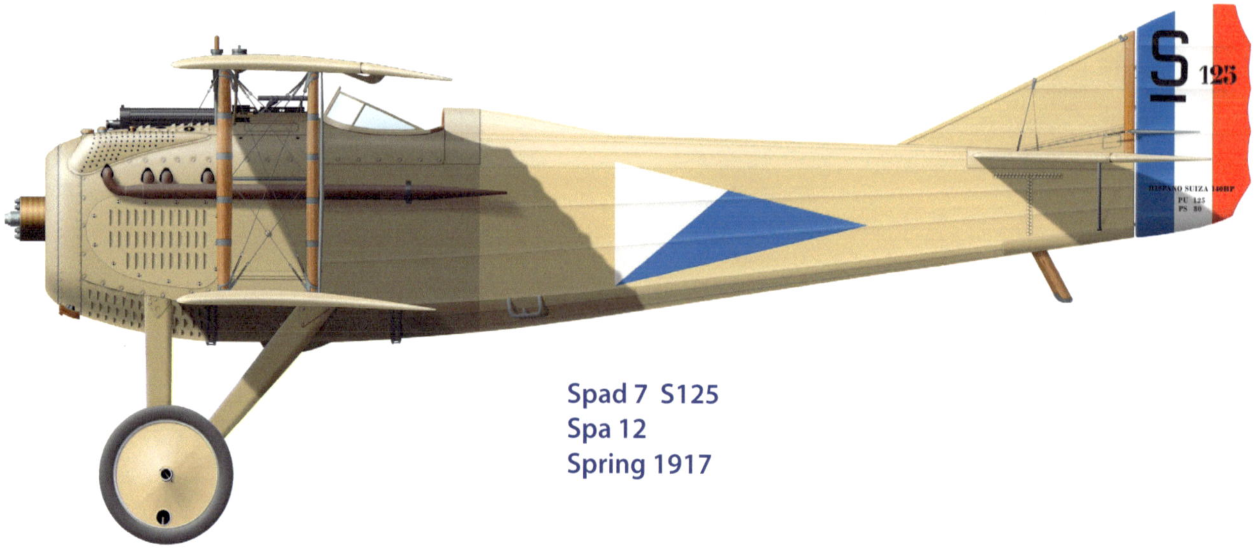

Spad 7 S125
Spa 12
Spring 1917

Charles Eugene Jules Marie Nungesser
(1892–1927)
Lieutenant
N 65, Spa 65, V 106, V 116
43 victories

Voisin LAS 3 V647
VB106
July 1915

Voisin LAS 3 V831
VB106
Summer 1915

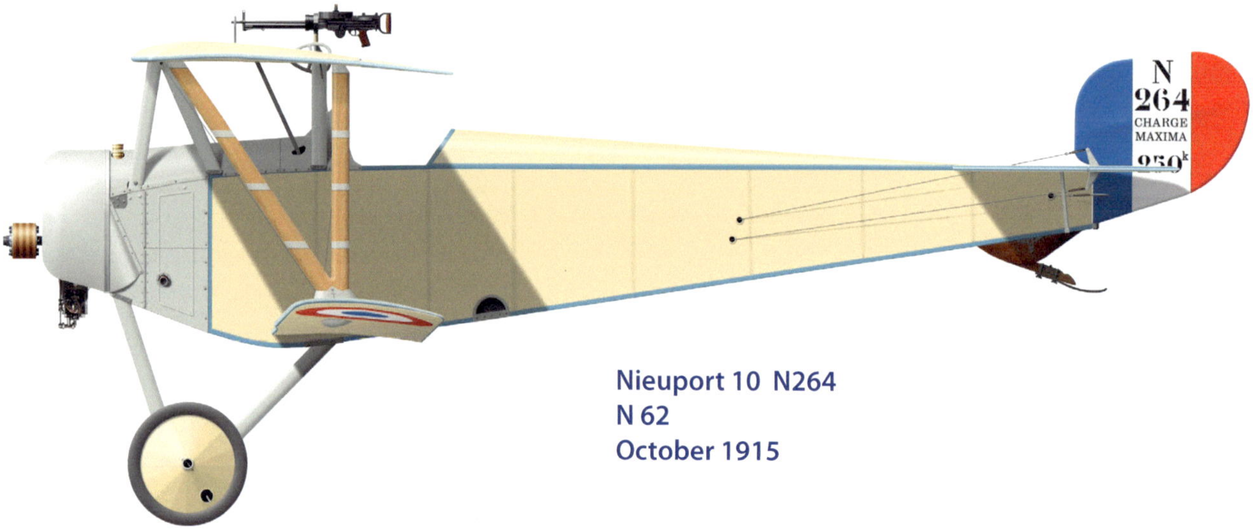

Nieuport 10 N264
N 62
October 1915

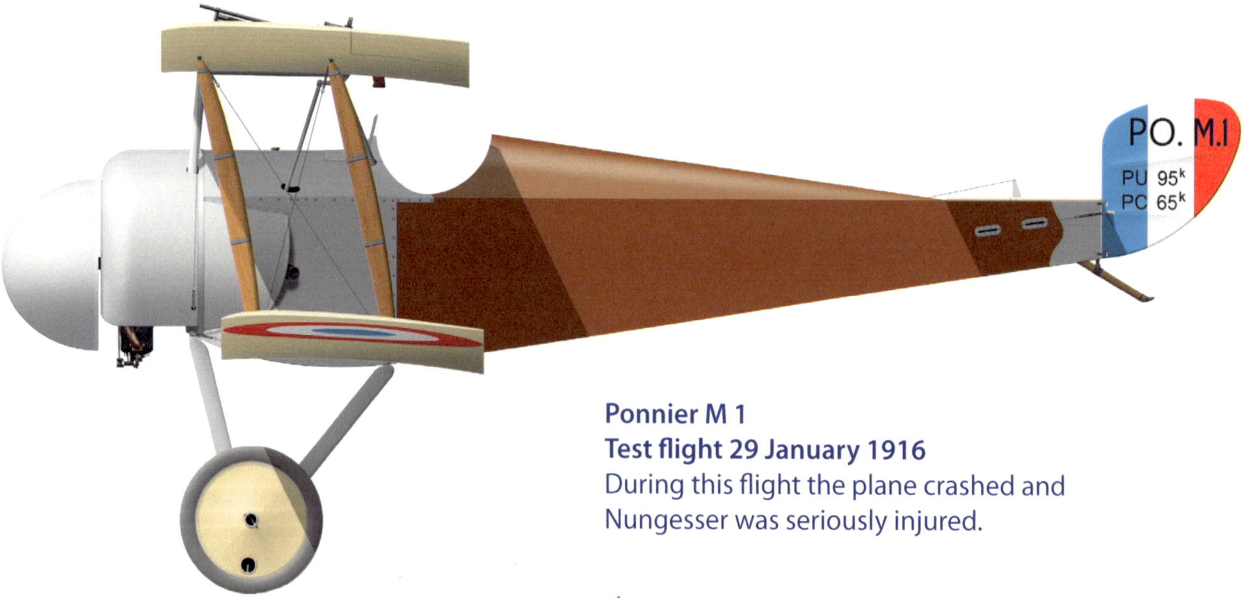

Ponnier M 1
Test flight 29 January 1916
During this flight the plane crashed and Nungesser was seriously injured.

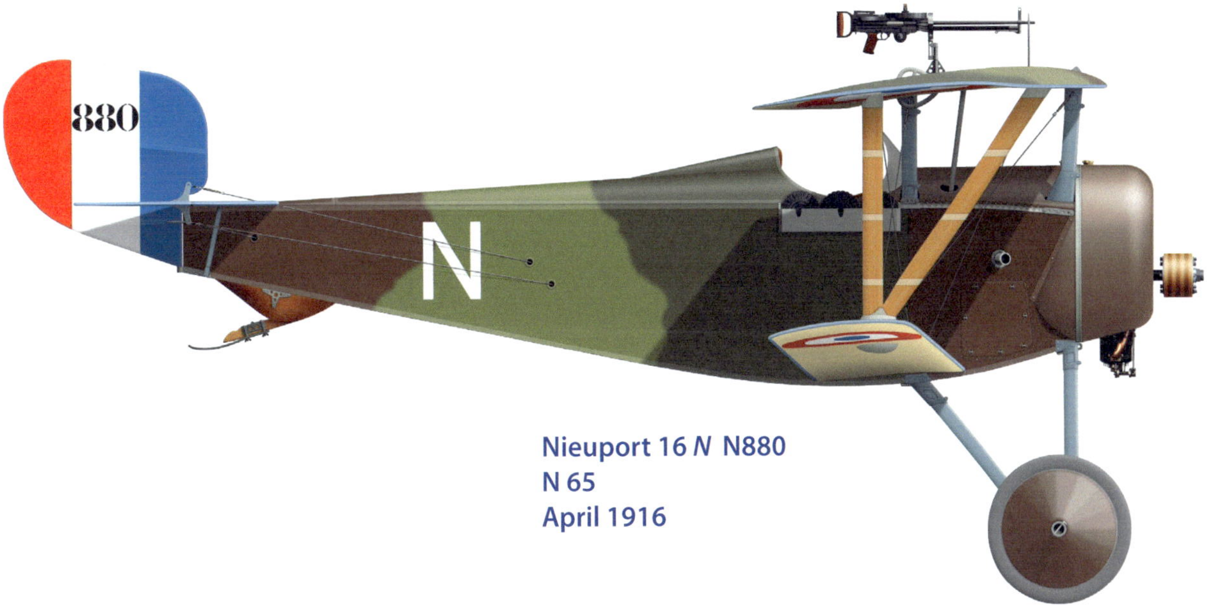

Nieuport 16 *N* N880
N 65
April 1916

Nieuport 16 *69 M* N938
N 65
May 1916
During front-line tests of the rockets Nungesser flew this airplane on May 22.

Nieuport 16 *N* N1139
N 65
May 1916

Nieuport 17 N1490
N 65
July 1916

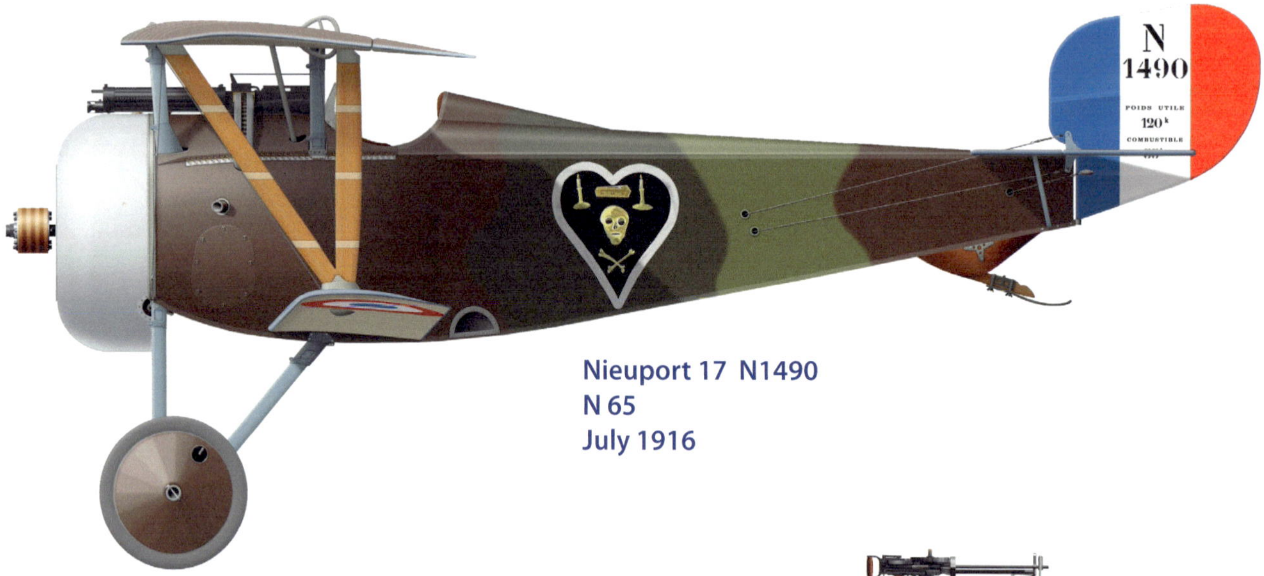

Nieuport 17 N1490
N 65
July 1916

Nieuport 17 N1571
N 65
Late summer 1916

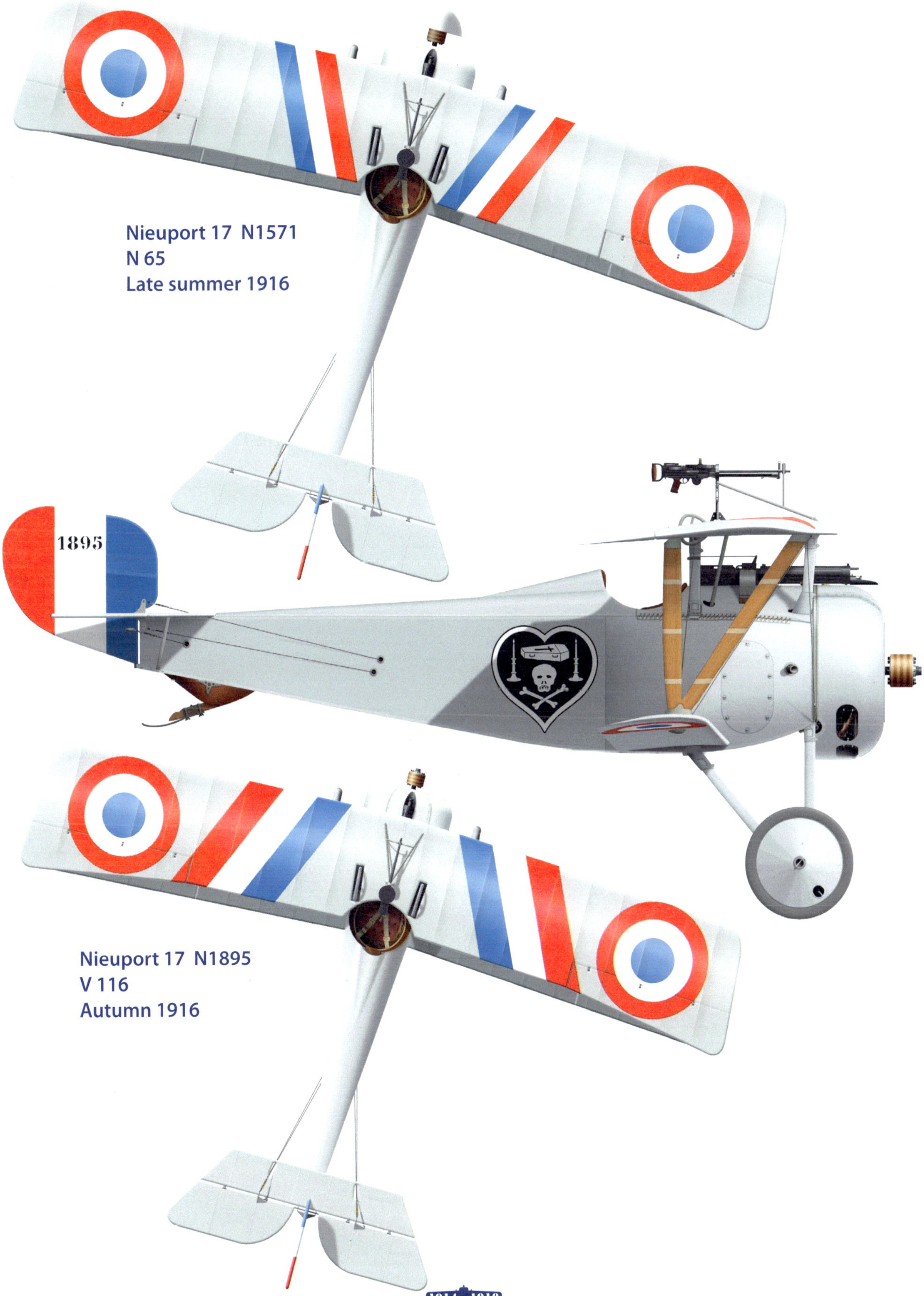

Nieuport 17 N1571
N 65
Late summer 1916

Nieuport 17 N1895
V 116
Autumn 1916

Nieuport 17*bis* N1895
N 65
Spring 1917

Nieuport 17*bis* N1895
N 65
Spring 1917

Nieuport 25 N5324
V 116
Summer 1917
The serial of this plane was later changed to N1895.

Nieuport 25 N5324
V 116
Summer 1917

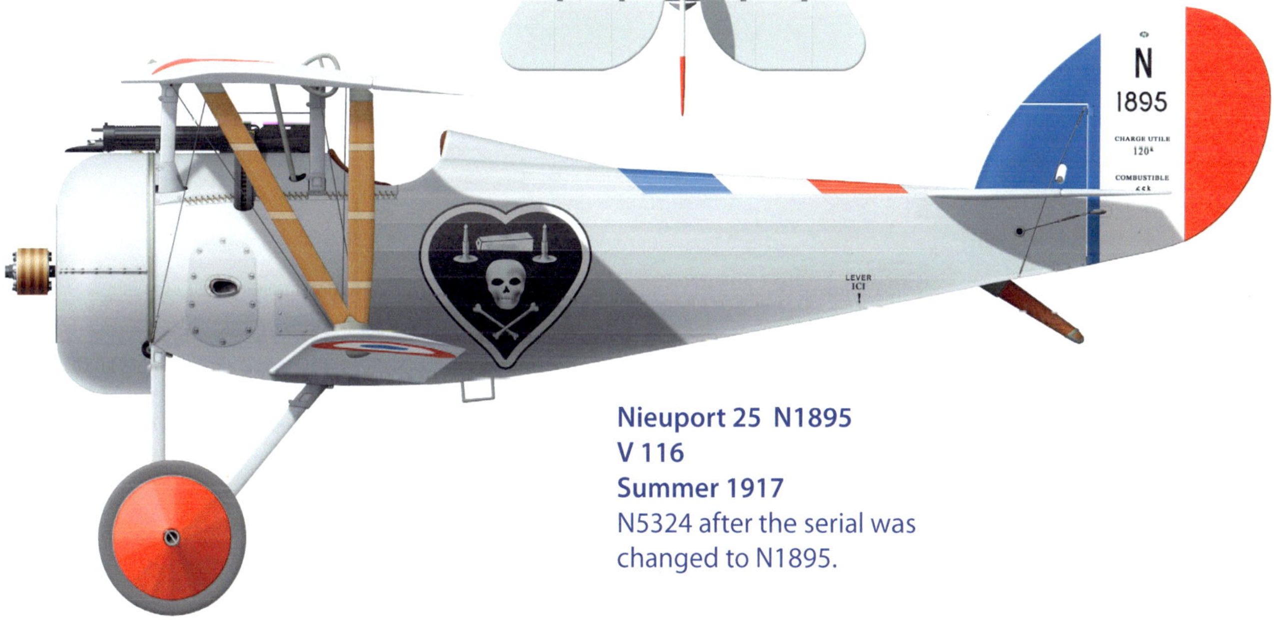

Nieuport 25 N1895
V 116
Summer 1917
N5324 after the serial was changed to N1895.

Nieuport 28 N6187
August 1917
Demonstration flights for Americans.

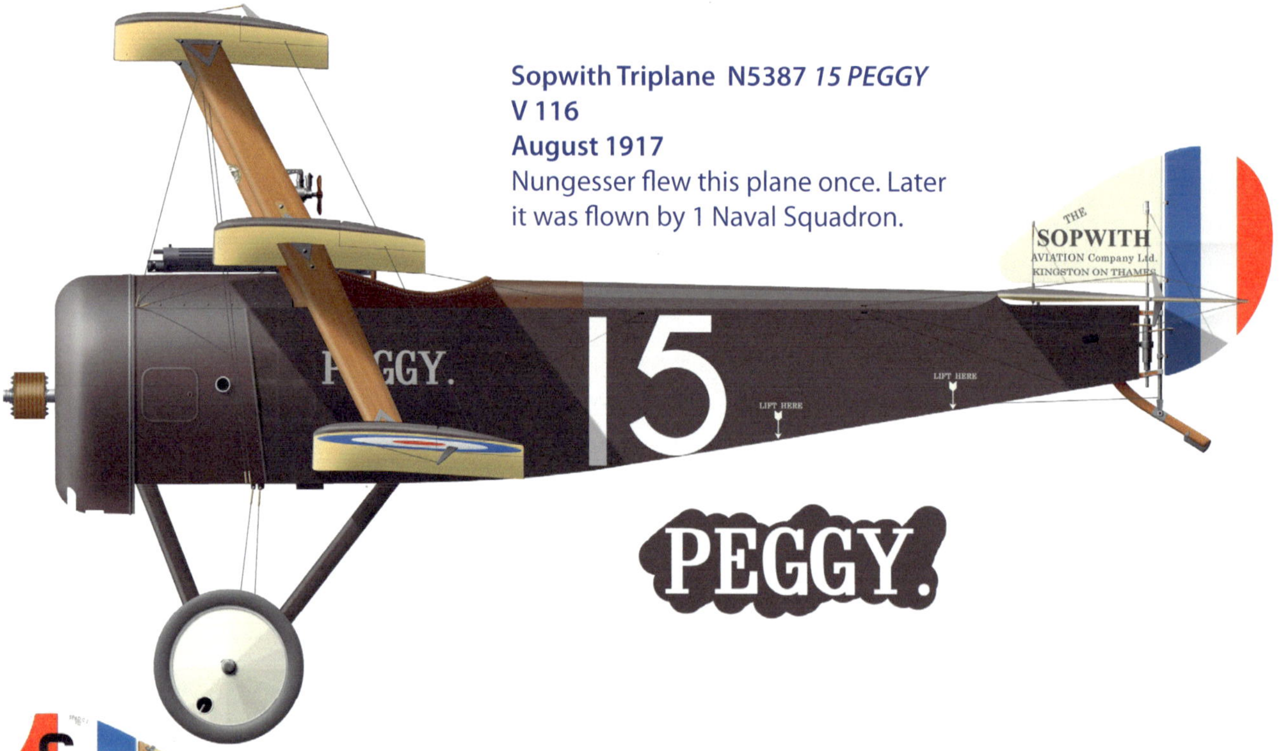

Sopwith Triplane N5387 *15 PEGGY*
V 116
August 1917
Nungesser flew this plane once. Later it was flown by 1 Naval Squadron.

Spad 7 S421
Spa 65
September 1917
This was probably the personal plane of the commander of Spa 65, *Capt*. Barthélémy.

Spad 12 S451
Spa 65
Summer 1918

Spad 13 *Lt Verdier* S10039
Spa 65
August 1918

Jacques Toussaint François Ortoli
(1895–1947)
Lieutenant
MF 8, N 31, Spa 31, N 77, Spa 57
11 victories

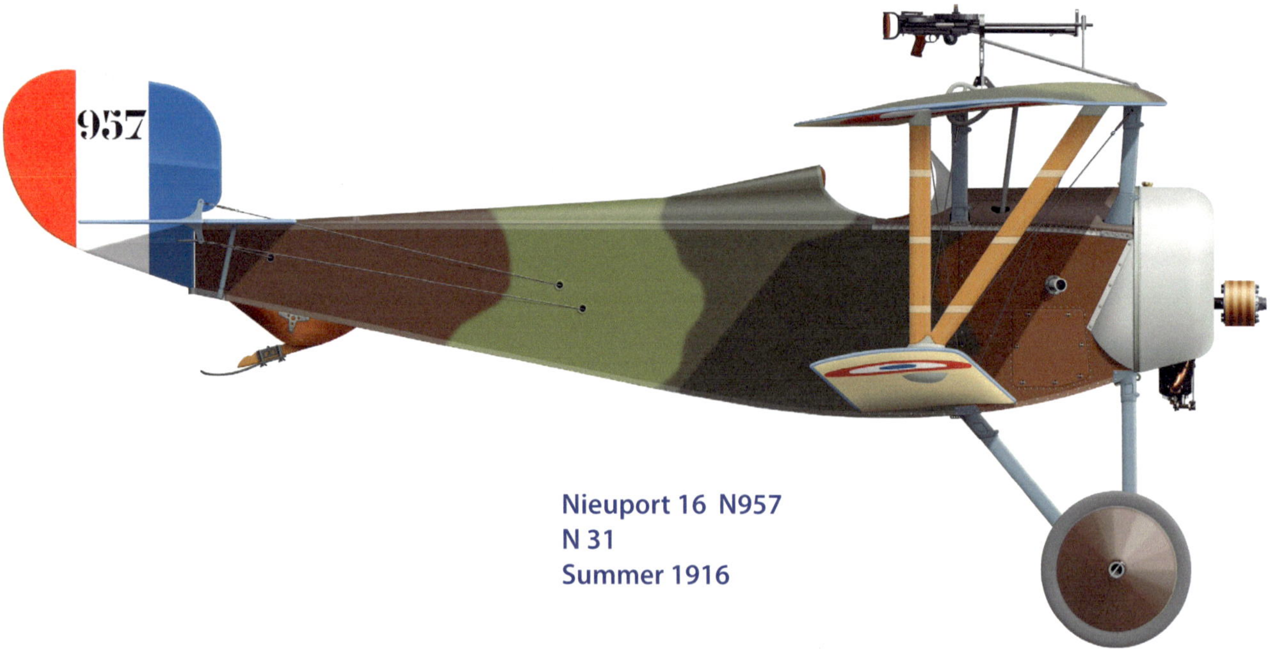

Nieuport 16 N957
N 31
Summer 1916

Spad 7
N 31
Spring 1917

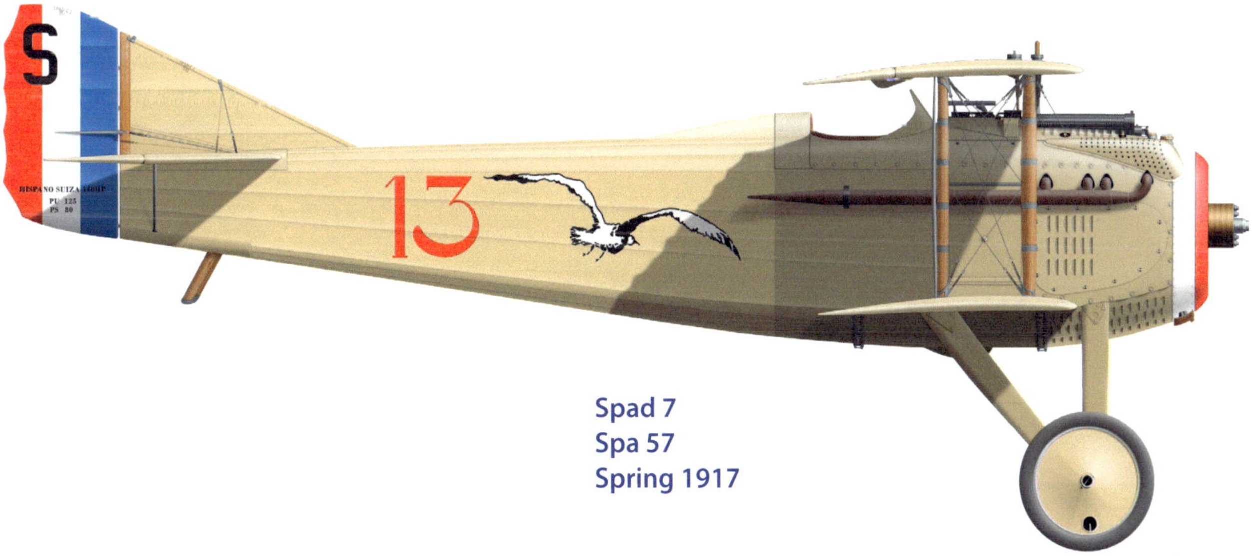

Spad 7
Spa 57
Spring 1917

Antoine Joseph Henri Louis Paillard
(1897–1931)
Sous-Lieutenant
Sop 111, BR 132
5 victories

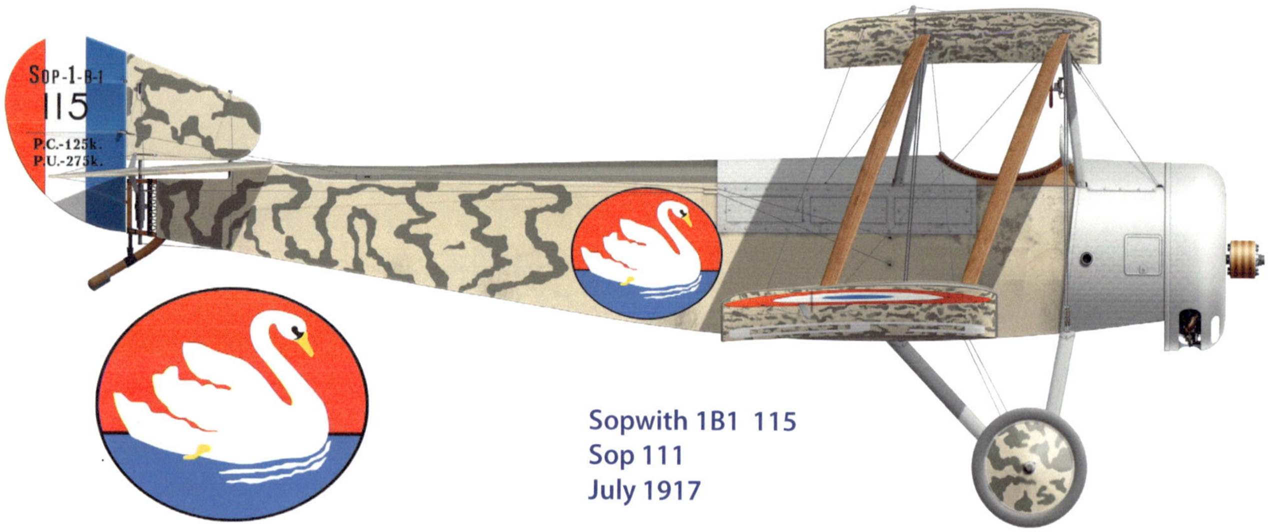

Sopwith 1B1 115
Sop 111
July 1917

Edwin Parsons
(1892–1968)
Adjudant
Spa 3
8 victories

Spad 13 S3831
Spa 3
September–October 1918

Adolphe Celestin Pègoud
(1889–1915)
Sous-Lieutenant
MF 25, MS 37, MS 49
6 victories

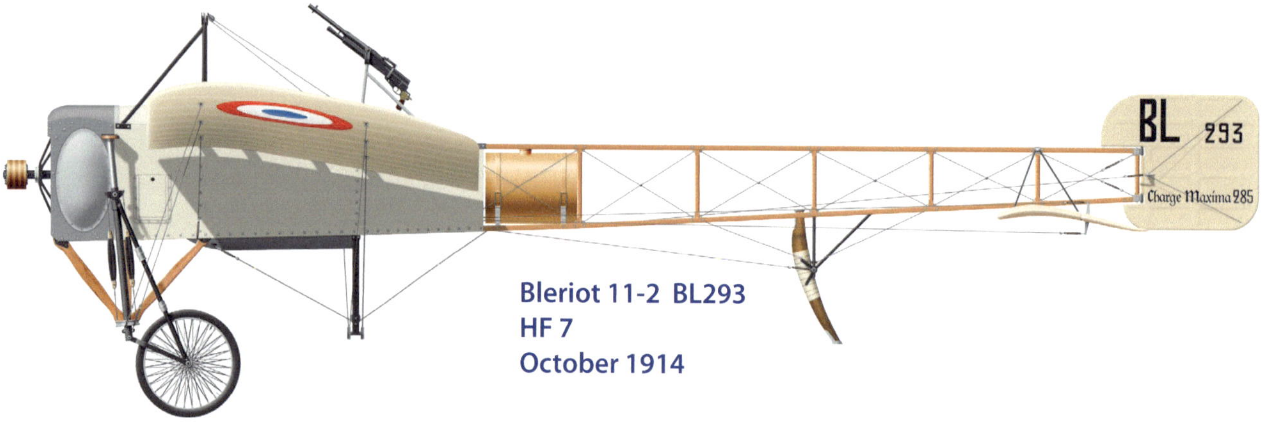

Bleriot 11-2 BL293
HF 7
October 1914

Morane-Saulnier L MS230
MF 25
February 1915

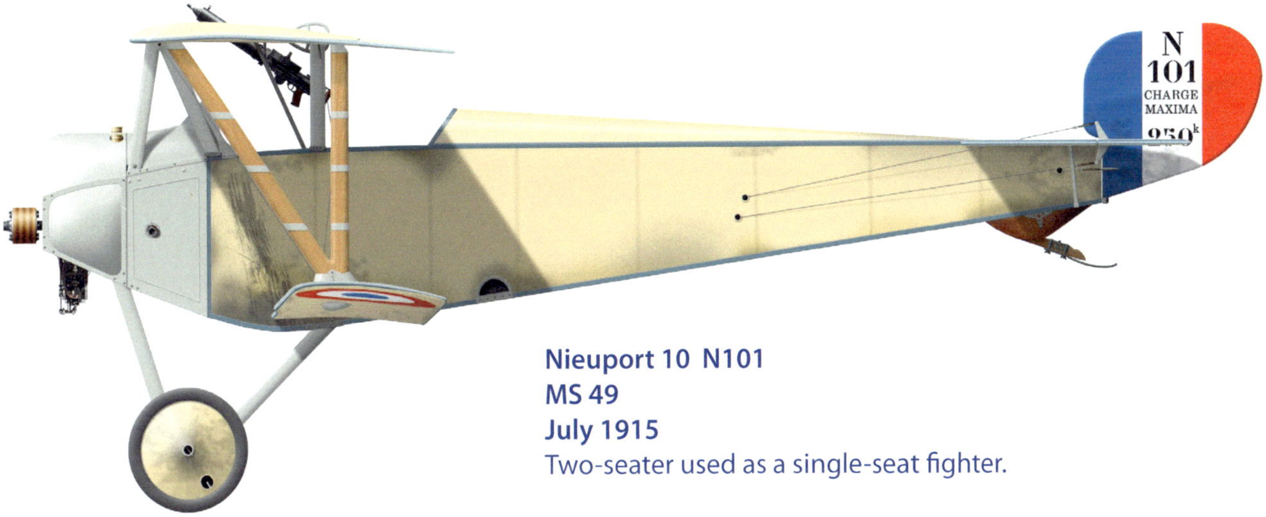

Nieuport 10 N101
MS 49
July 1915
Two-seater used as a single-seat fighter.

Nieuport 10 N144
MS 49
August 1915
Single-seat fighter.

Nieuport 10 N210
MS 49
August 1915

René Alphonse Pelissier
(1889–1969)
Sous-Lieutenant
C 53, C 202, N 155, Spa 175
6 victories

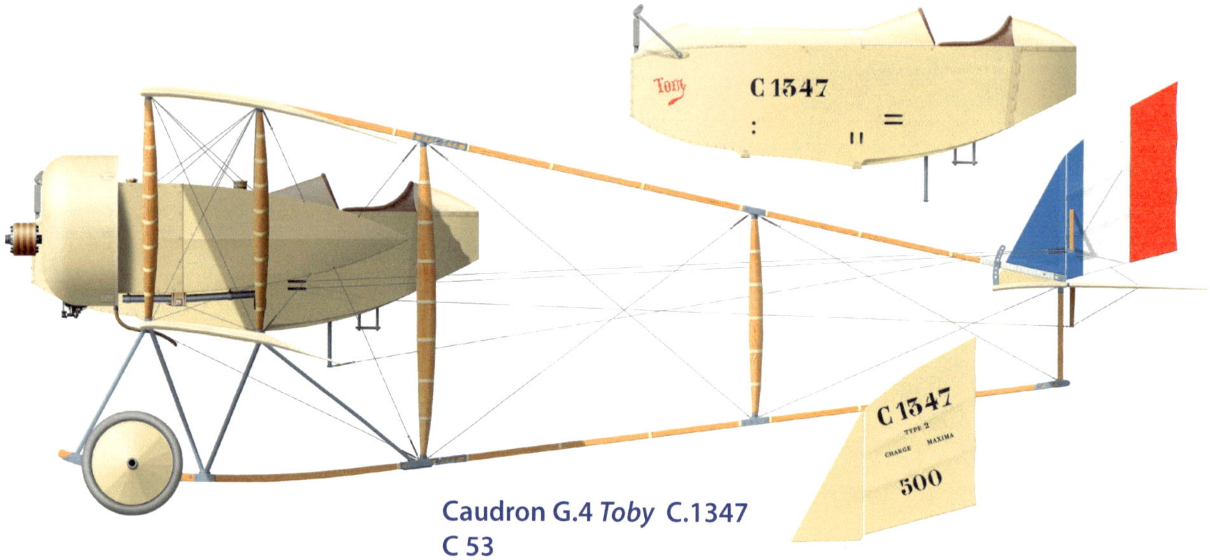

Caudron G.4 *Toby* C.1347
C 53
March 1916

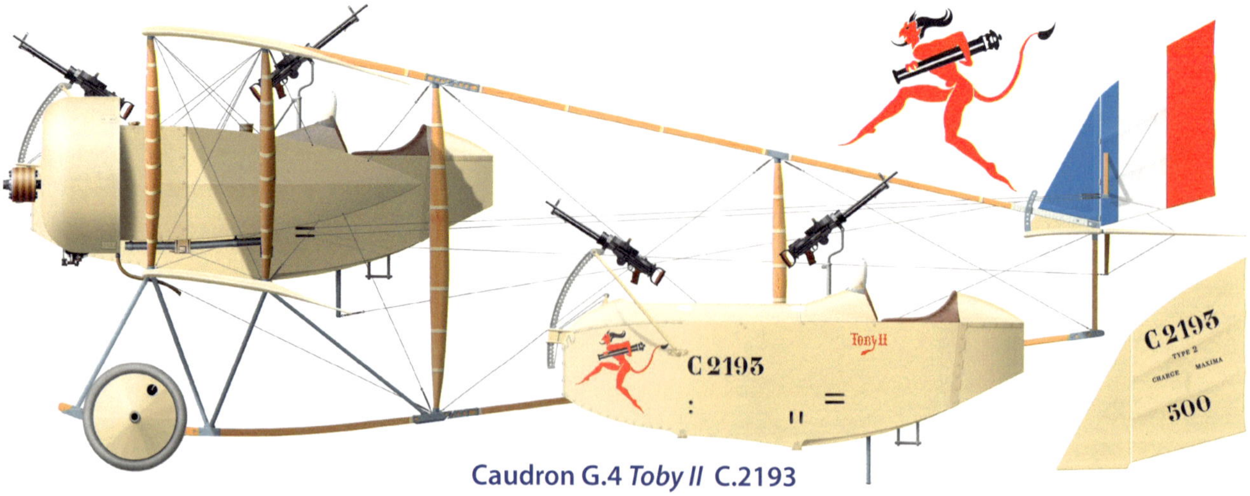

Caudron G.4 *Toby II* C.2193
C 202
June 1916

Nieuport 21 1687
Military Flying School Tour
Spring 1917

Spad 7 *Toby III* S3119
Spa 155
February–May 1918

Spad 7 S7221
Spa 155
June–September 1918

Georges Pelletier d'Oisy

(1892–1953)
Capitaine
HF 19, MS 12, N 69, Spa 69
5 victories

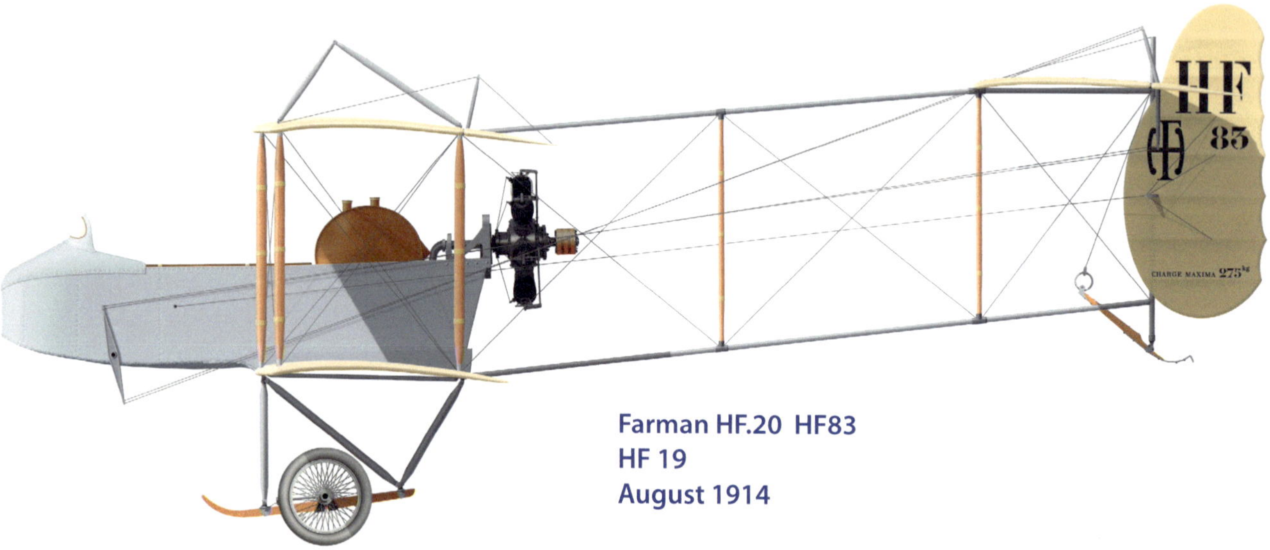

Farman HF.20 HF83
HF 19
August 1914

Morane-Saulnier L *Le Pilou, Pilou*
MS 12
Winter 1915–1916

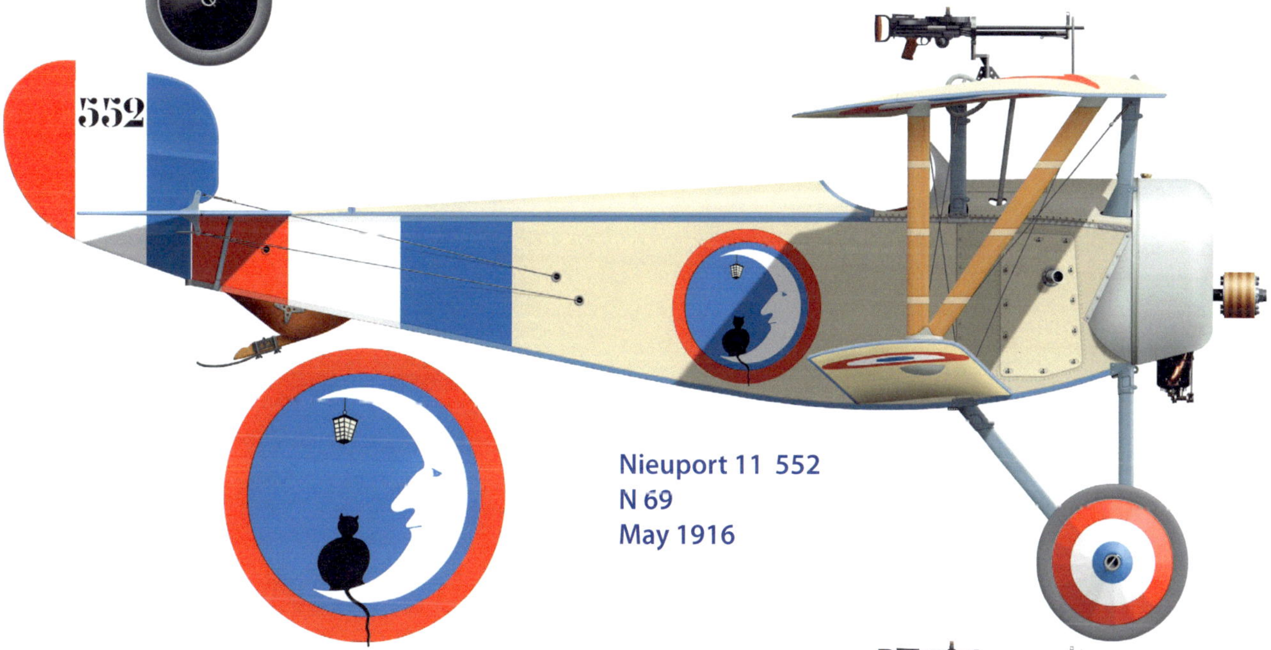

Nieuport 11 552
N 69
May 1916

Nieuport 16 *Pilou, Pilou VI* N825
N 69
Autumn 1916

Morane-Saulnier P
Pilou, Pilou VIII N825
N 69
Winter 1916–1917

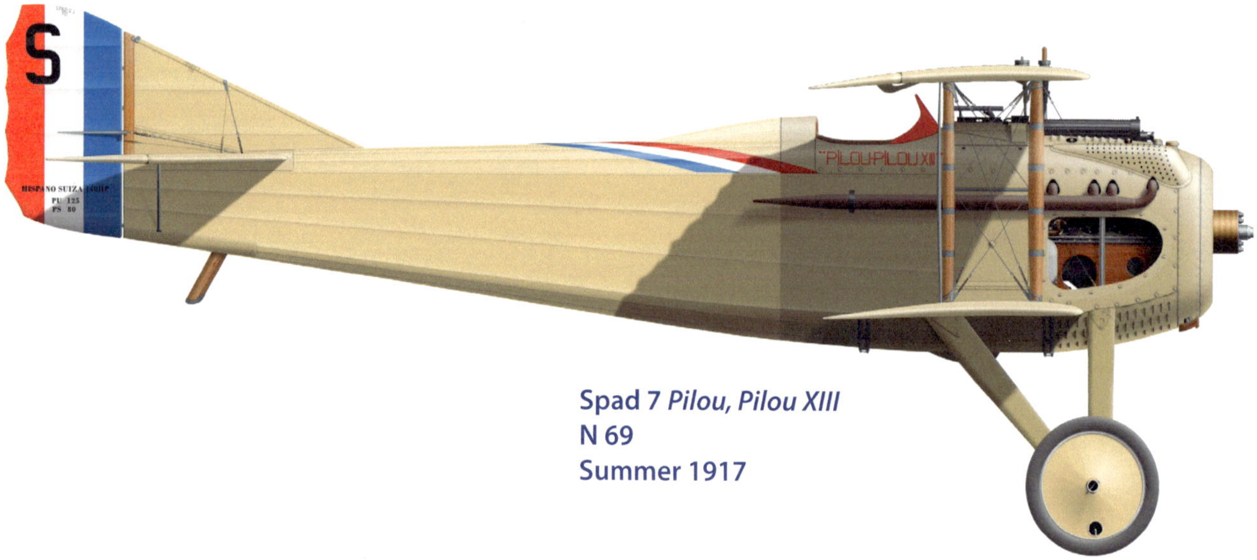

Spad 7 *Pilou, Pilou XIII*
N 69
Summer 1917

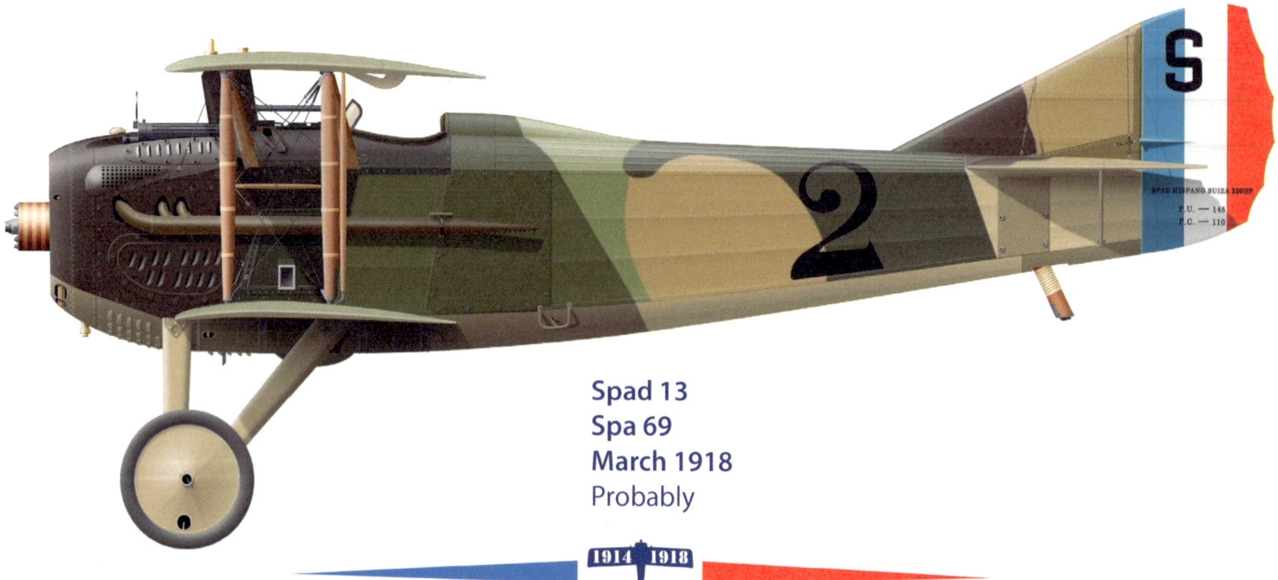

Spad 13
Spa 69
March 1918
Probably

Pierre Pendaries
(1894–1987)
Sous-Lieutenant
N 69, Spa 67
7 victories

Nieuport 17 N1199
N 69
Late 1916

Henri Albert Péronneau
(1890–1960)
Sous-Lieutenant
N49, N65, N81, Spa81
9 victories

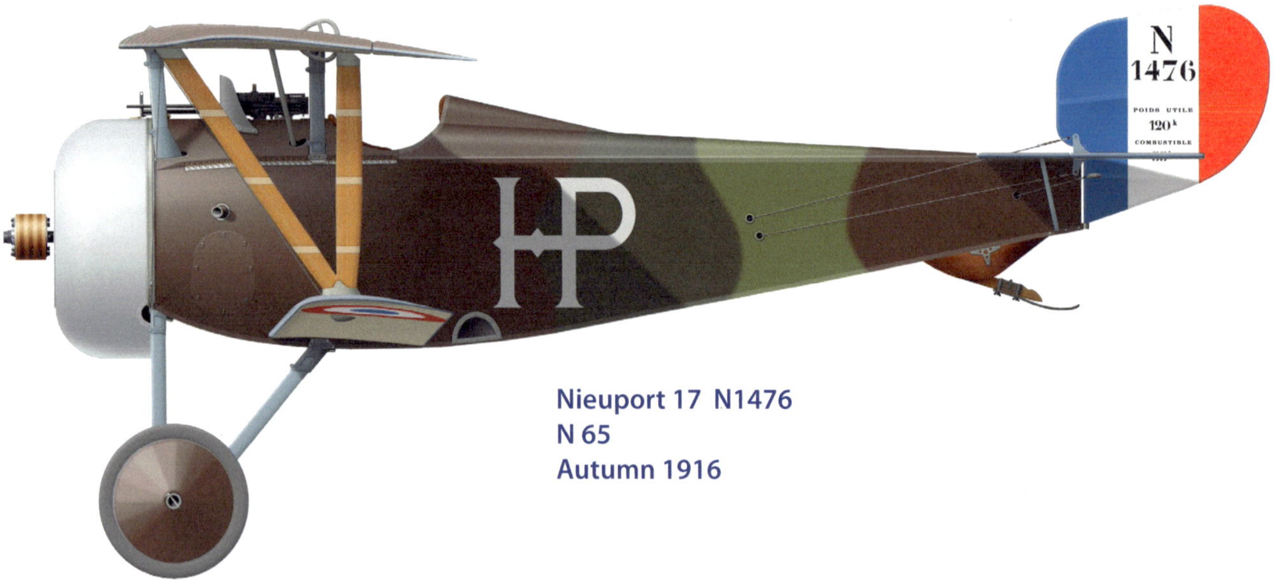

Nieuport 17 N1476
N 65
Autumn 1916

Paul Armand Petit

(1890–1918)
Adjudant
N 154, Spa 154
7 victories

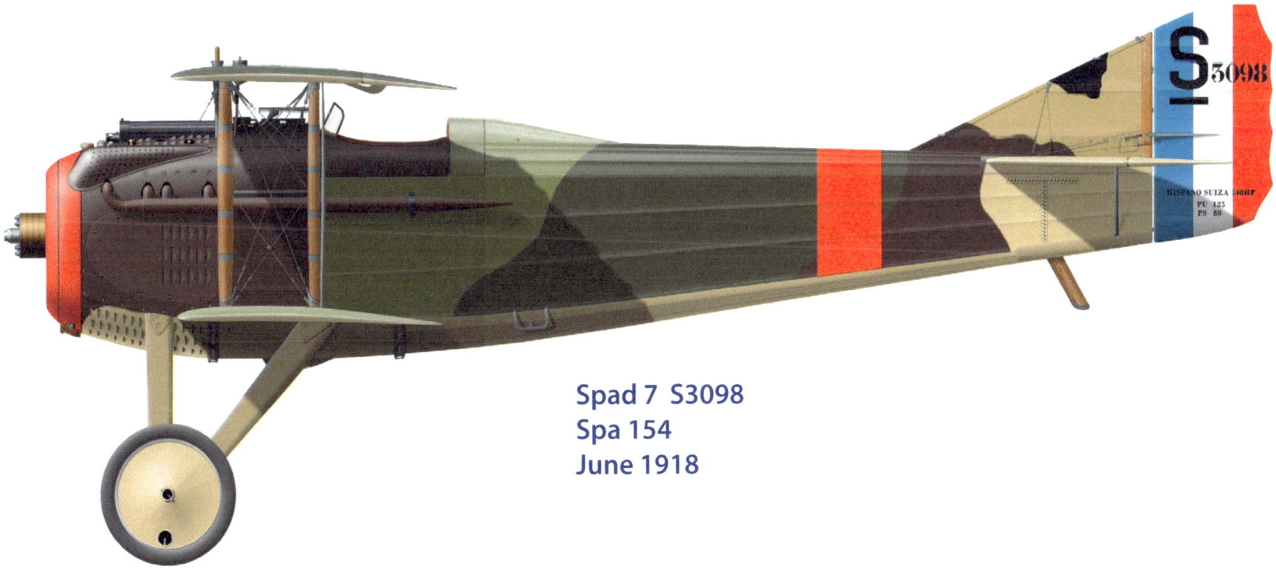

Spad 7 S3098
Spa 154
June 1918

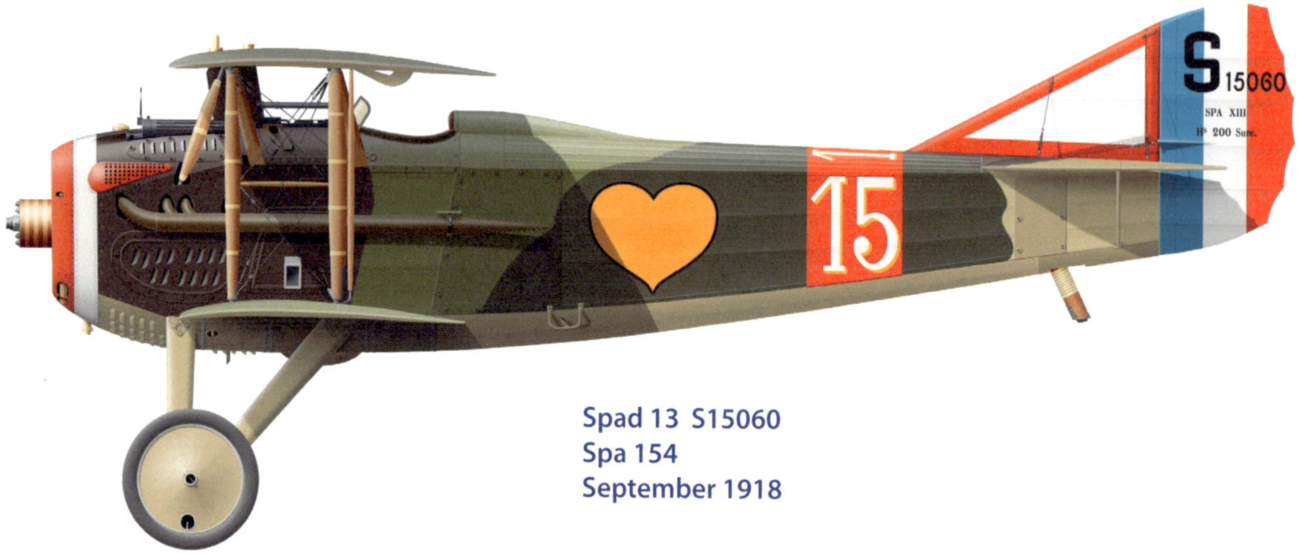

Spad 13 S15060
Spa 154
September 1918

Edmond Jacques Marcel Pilon
(1891–1921)
Adjudant
MF 20, N 102, N 82, N 57, Spa 82, Spa 67, Spa 98
8 victories

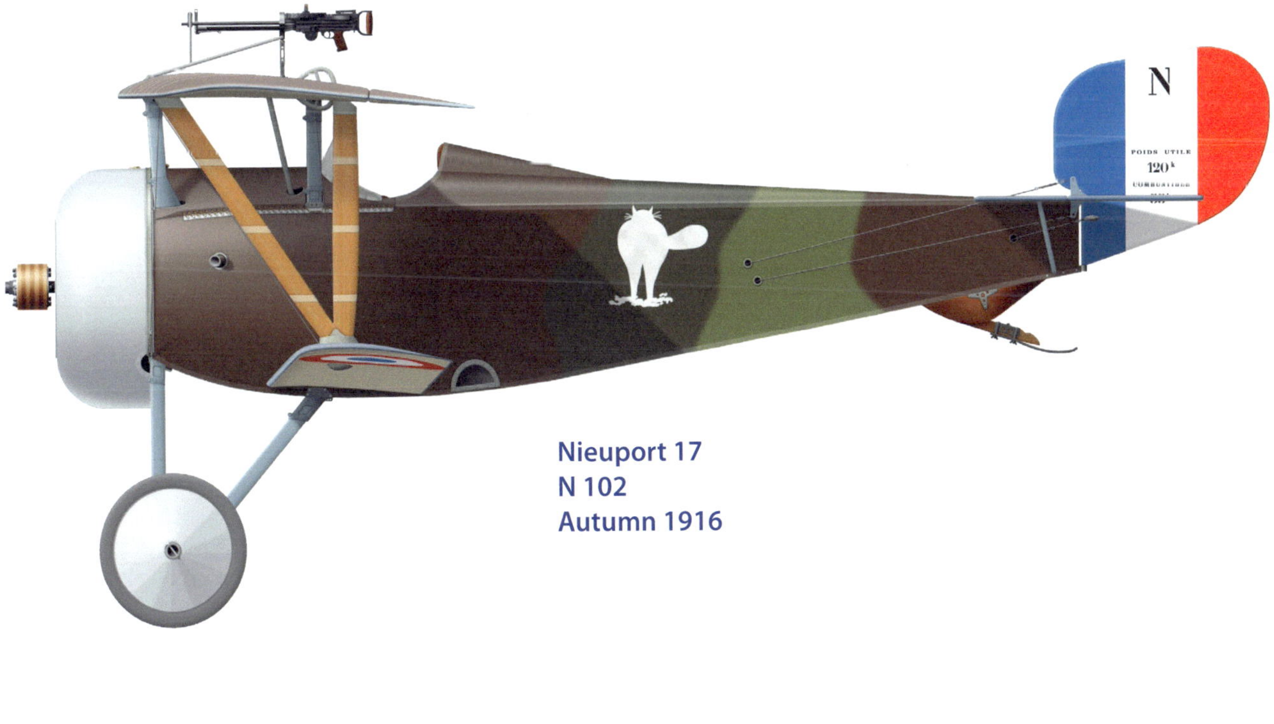

Nieuport 17
N 102
Autumn 1916

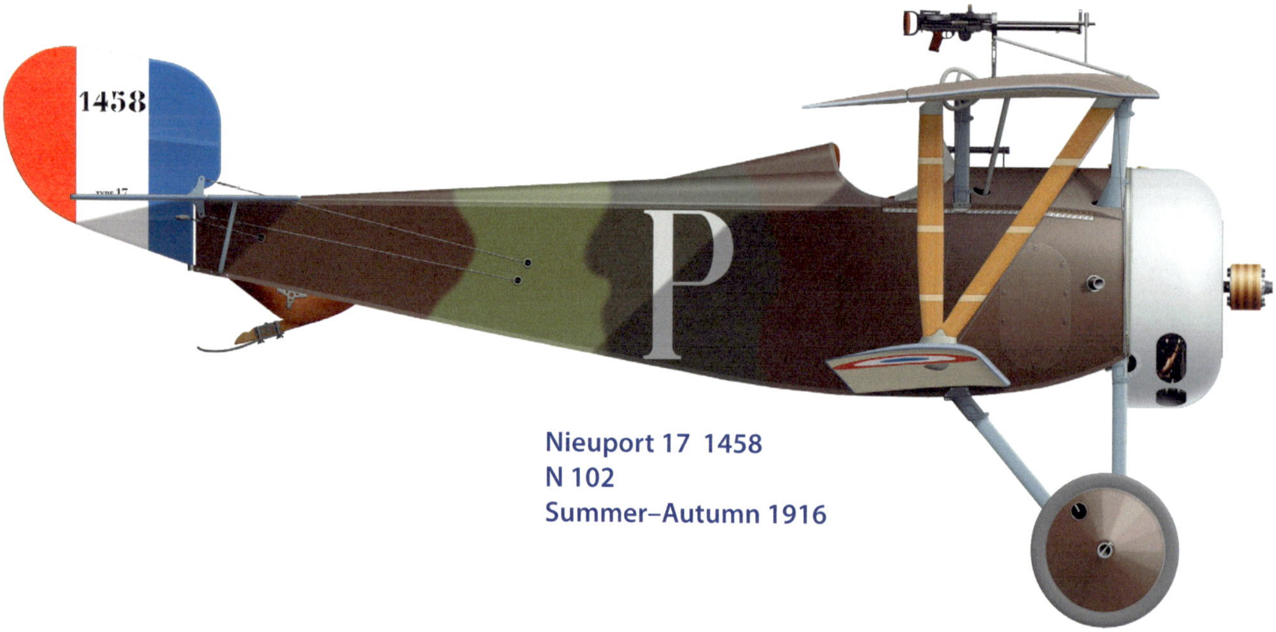

Nieuport 17 1458
N 102
Summer–Autumn 1916

Spad 7 S351
Spa 82
September 1917

Armand Pinsard
(1887–1953)
Capitaine
MS 23, N 26, N 78, SPA 23
27 victories

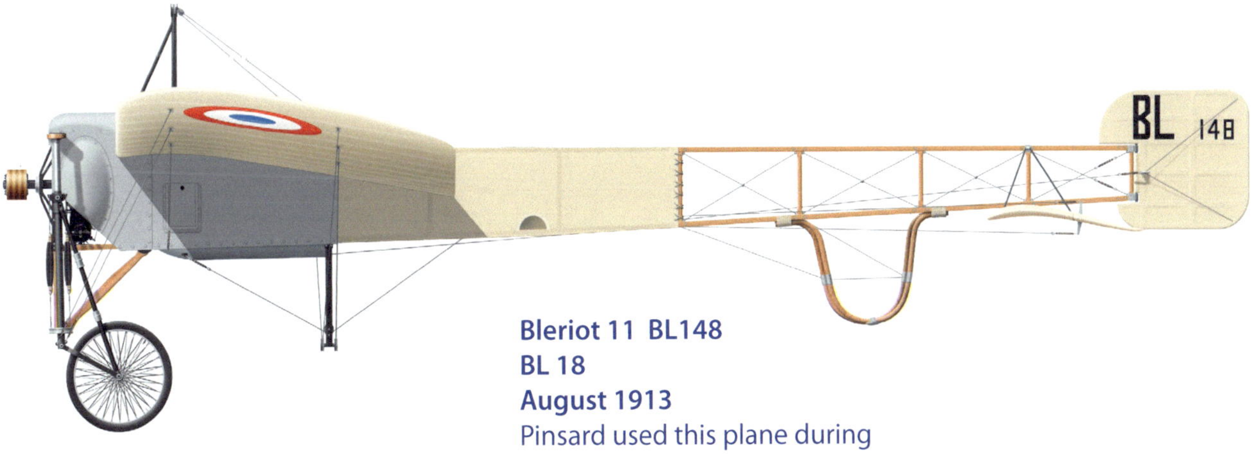

Bleriot 11 BL148
BL 18
August 1913
Pinsard used this plane during army maneuvers and was awarded the Military Medal for that.

Morane-Saulnier L MS23
MS 23
September 1914

Nieuport 16 N1377
N 3
Summer 1916

Spad 7 S122
N 26
Summer 1916
Before painting.

Spad 7 S122
N 26
Summer 1916
After painting.

Spad 7 *Revanche* S147
N 78
Winter 1917

Spad 7 *Revanche IV* S122
N 78
March 1917

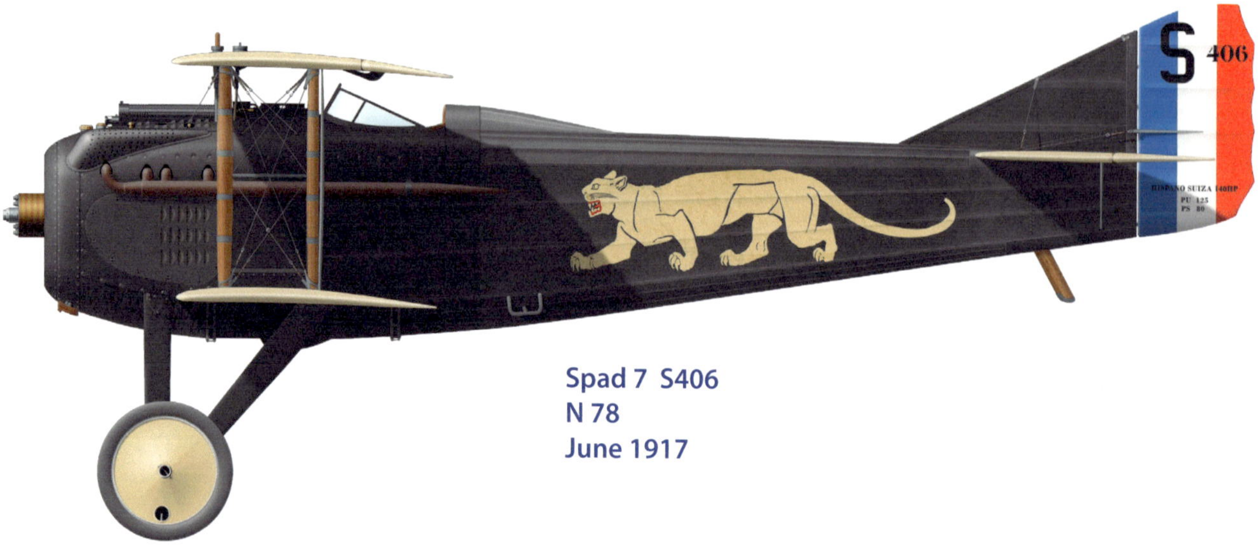

Spad 7 S406
N 78
June 1917

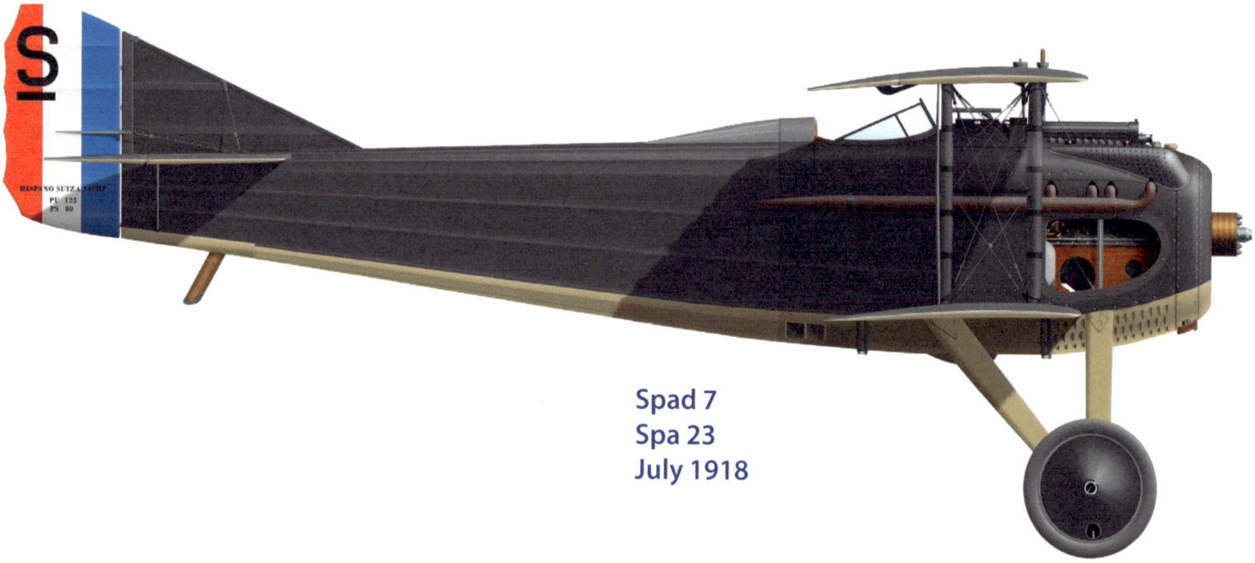

Spad 7
Spa 23
July 1918

Morane-Saulnier MS.406 C1 *Le Pirate* L-609 N1019
GC 21
May 1940

Constant Emmanuel Plessis
(1890–1930)
Sous-Lieutenant
N88, Spa65
6 victories

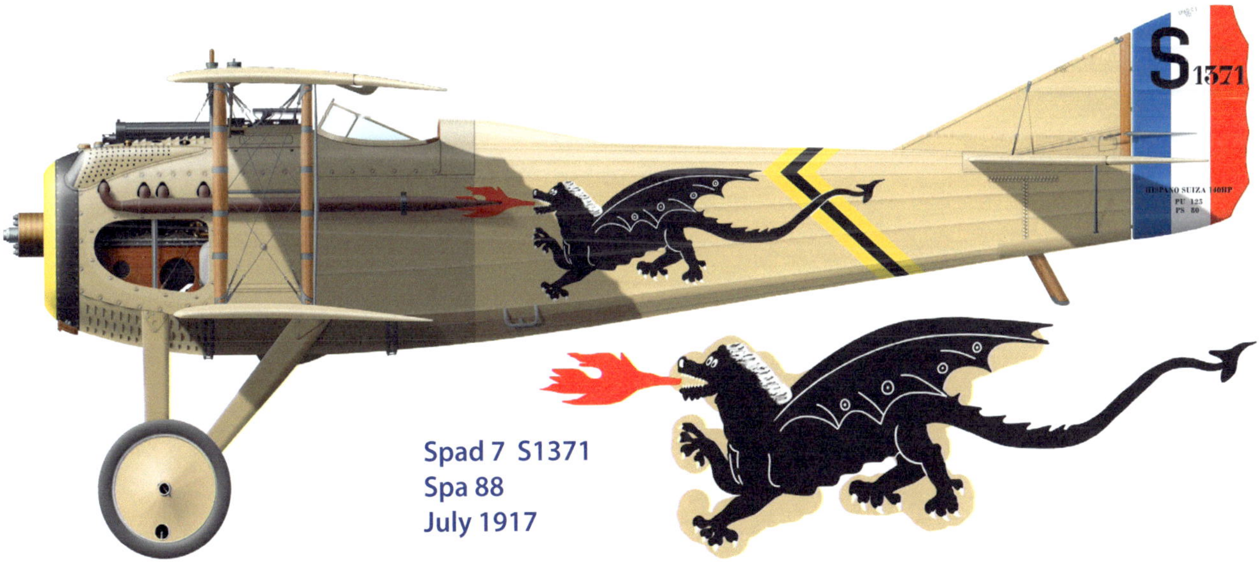

Spad 7 S1371
Spa 88
July 1917

Roger Poupon
(1888–1976)
Capitaine
N 37, Spa 37
8 victories

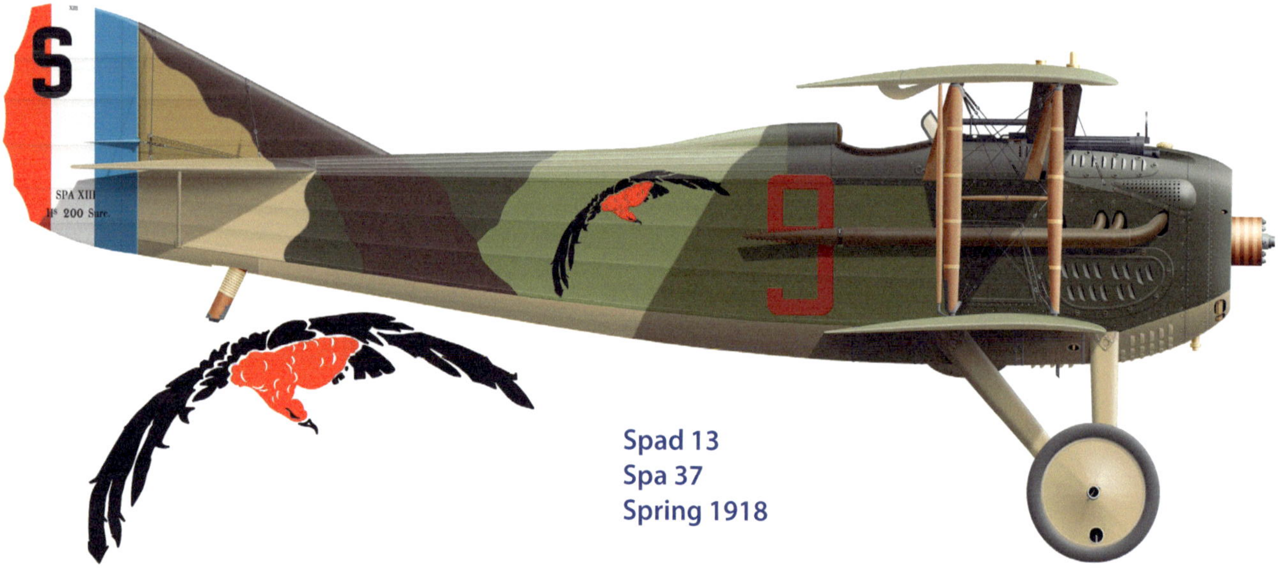

Spad 13
Spa 37
Spring 1918

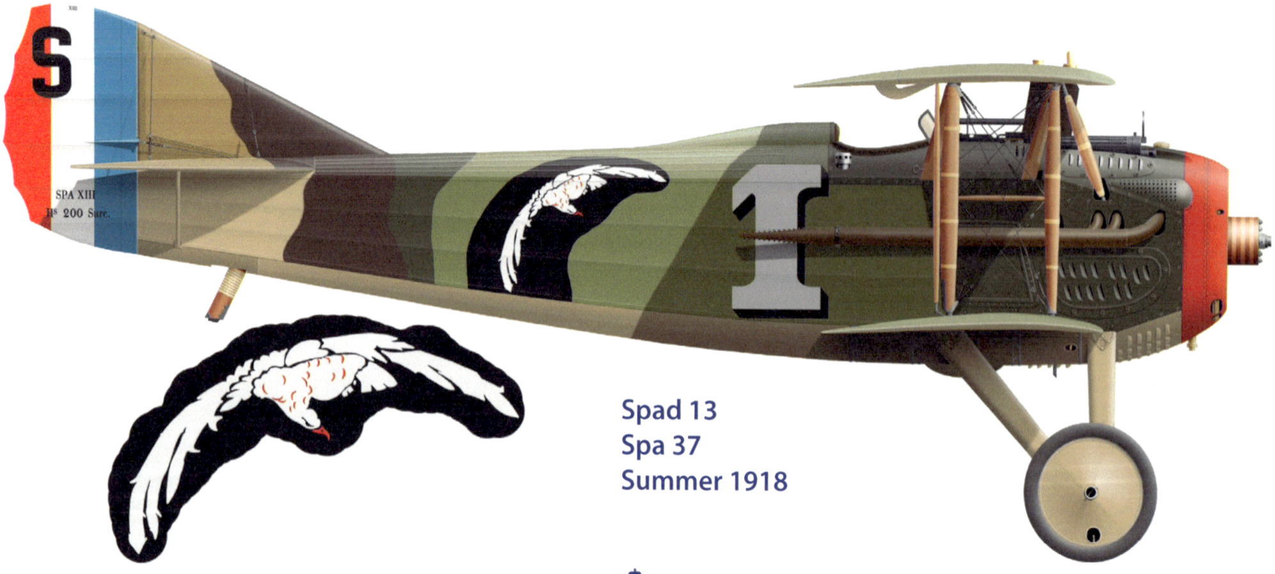

Spad 13
Spa 37
Summer 1918

Pulpe Eduard Martinovich (Edward Pulpe) (1880–1916)
Sous-Lieutenant
MS 23, 10-th IAO, 8-th IAO
5 victories

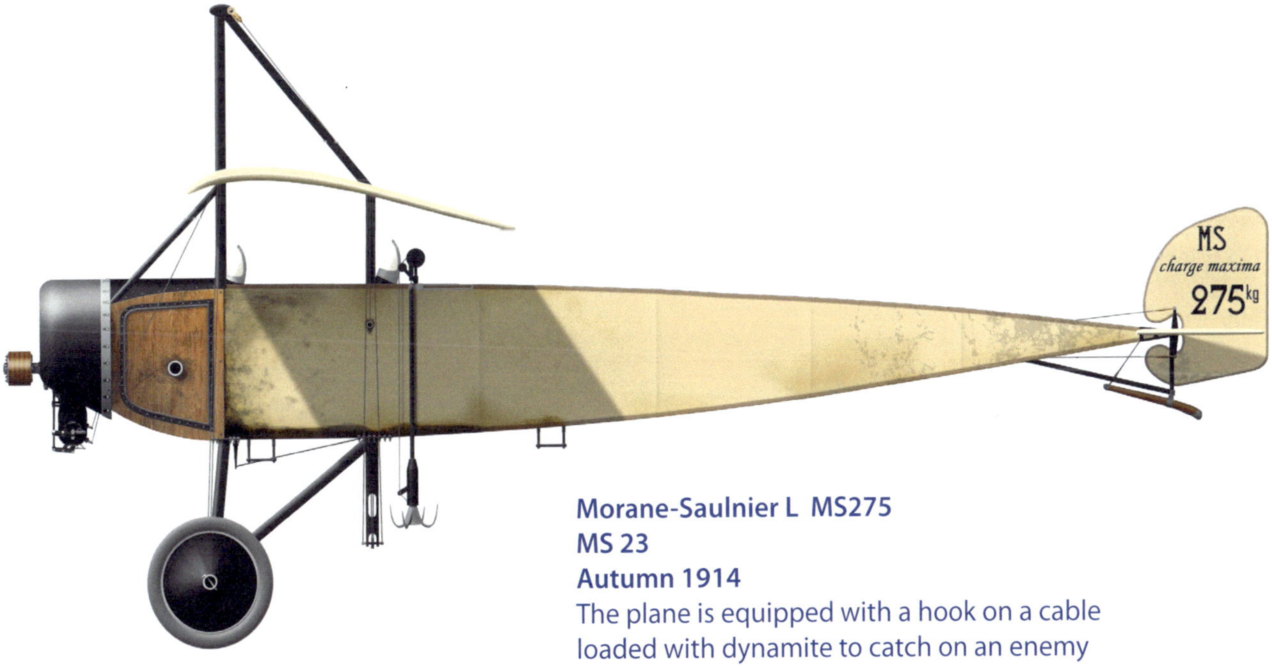

Morane-Saulnier L MS275
MS 23
Autumn 1914
The plane is equipped with a hook on a cable loaded with dynamite to catch on an enemy plane. This was very dangerous to Pulpe.

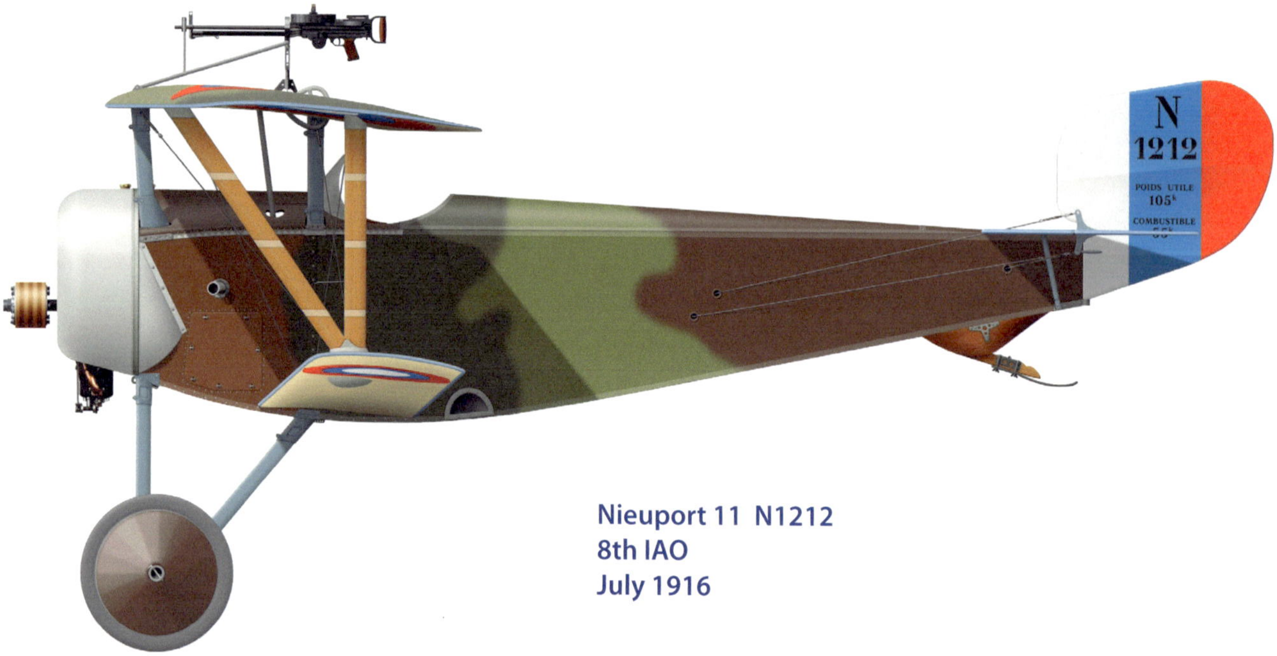

Nieuport 11 N1212
8th IAO
July 1916

Georges Raymond
(1887–1918)
Capitaine
N 3, Spa 3
5 victories

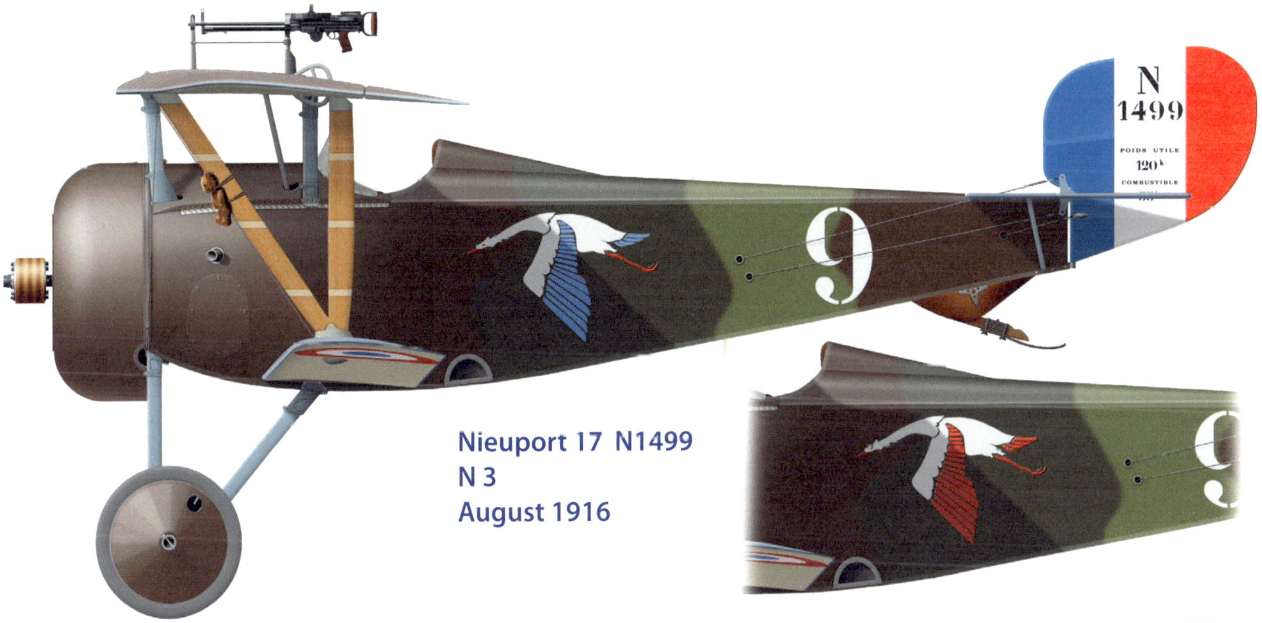

Nieuport 17 N1499
N 3
August 1916

Nieuport 17 *Ma Ninon* N1538
N 3
October 1916

Nieuport 17 N1538
N 3
October 1916

Spad 7 *Ma Ninion* S313
N 3
February–March 1917

Spad 7 *Ma Ninion* S1422
N 3
July 1917

Emile Julien Mathurin Regnier
(1894–1940)
Sous-Lieutenant
Spa 89
6 victories

Spad 7
Spa 89
Winter 1918

Victor François Marie Alexis Regnier

(1889–1971)
Lieutenant
N 112, Spa 112
5 victories

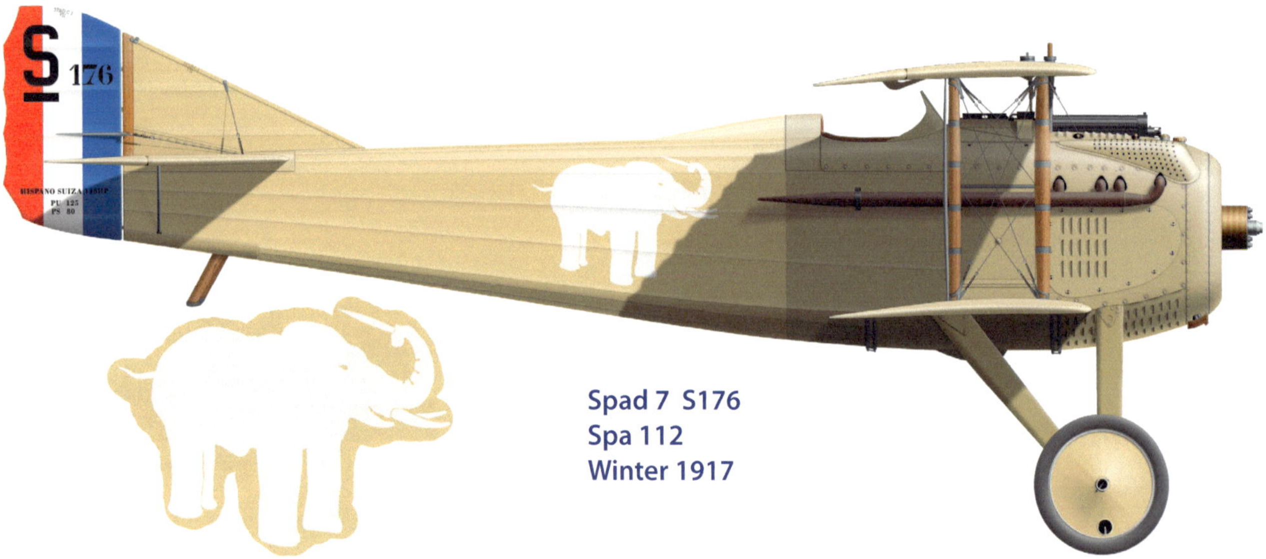

Spad 7 S176
Spa 112
Winter 1917

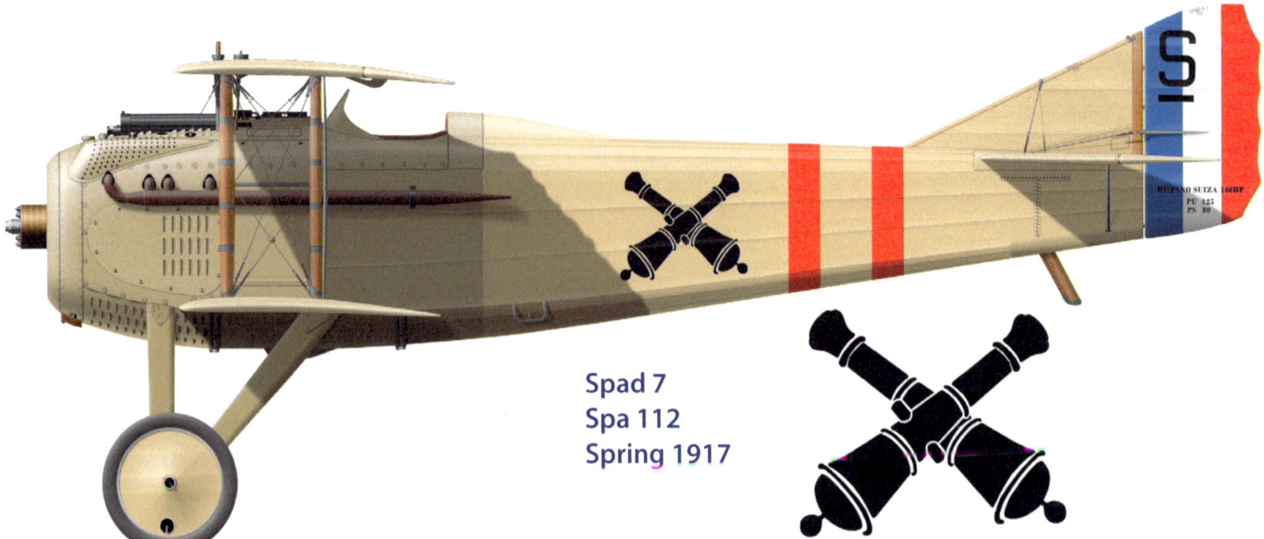

Spad 7
Spa 112
Spring 1917

Charles Alfred Revol-Tissot

(1892–1971)
Adjudant
BL 3, N 38
5 victories

Bleriot 11 BL206
BL 3
October 1914

Louis Risacher
(1894–1986)
Sous-Lieutenant
Spa 3, Spa 159
5 victories

Spad 7
Spa 3
September 1917

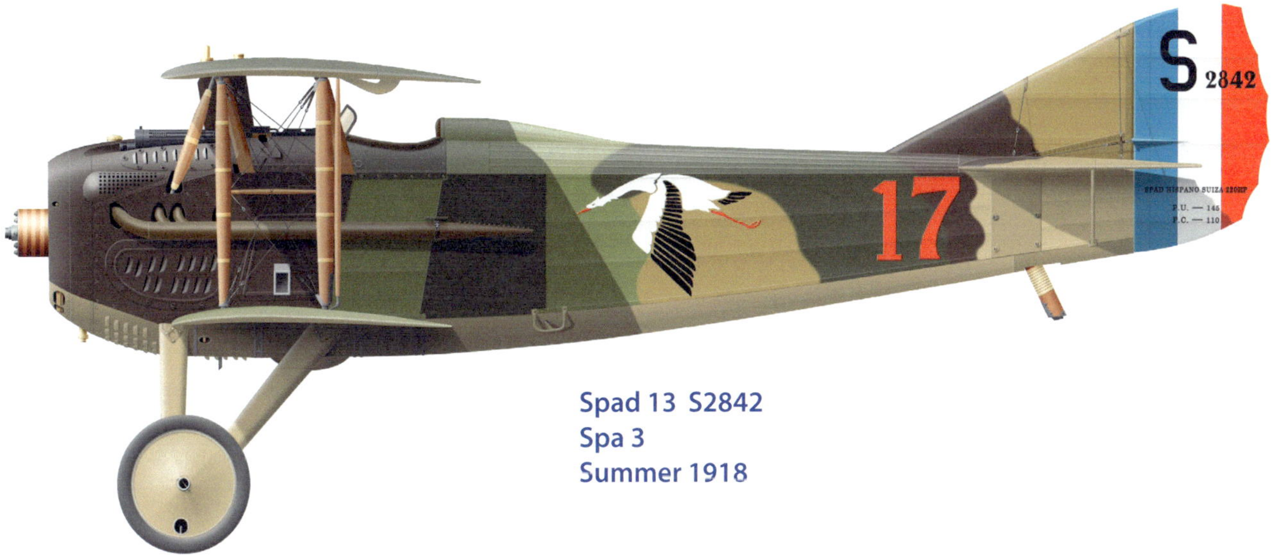

Spad 13 S2842
Spa 3
Summer 1918

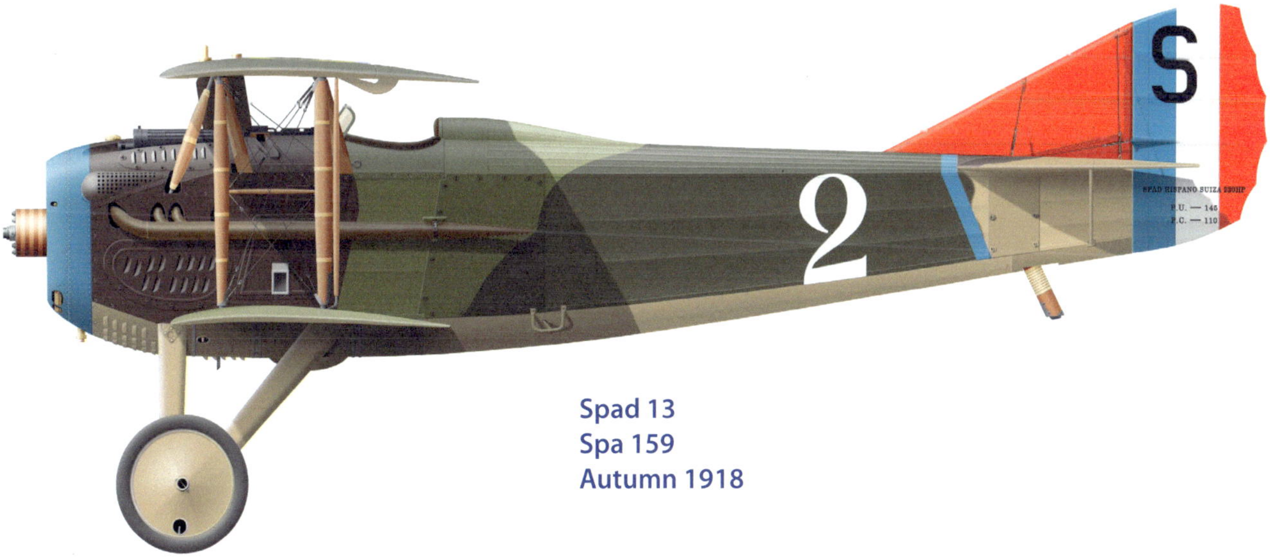

Spad 13
Spa 159
Autumn 1918

François Marie Laurent Viturien de Rochechouart, Marquis de Mortemart, Prince de Tonnay-Charente
(1881–1918) Sous-Lieutenant
N 23, Spa 23
7 victories

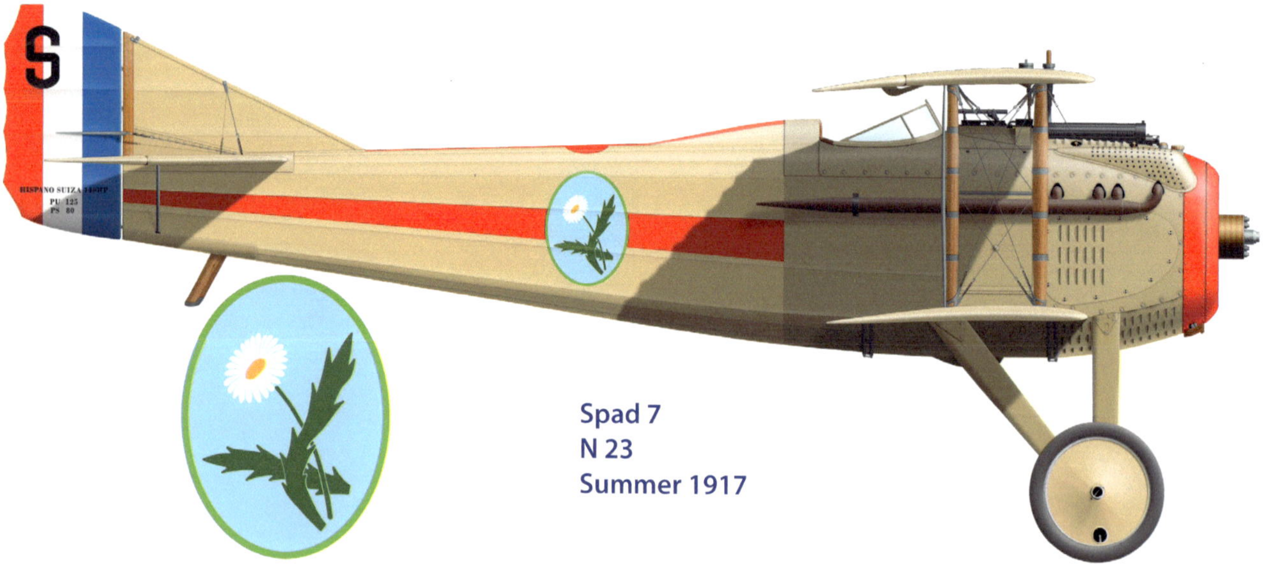

Spad 7
N 23
Summer 1917

Noel Hugues Anne Louis de Rochefort
(1887–1916)
Sous-Lieutenant
N 26, Spa 26
7 victories

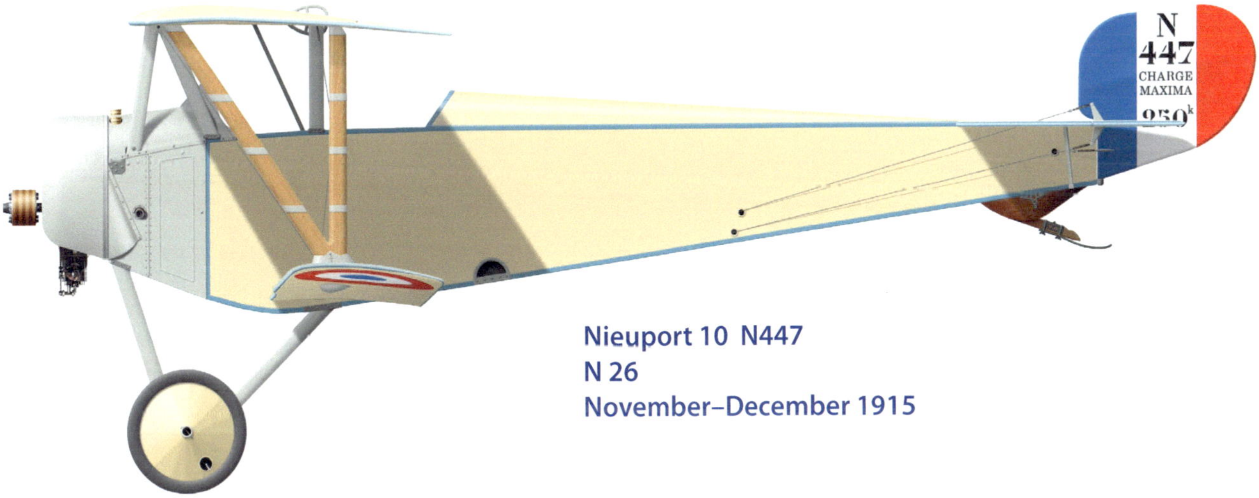

Nieuport 10 N447
N 26
November–December 1915

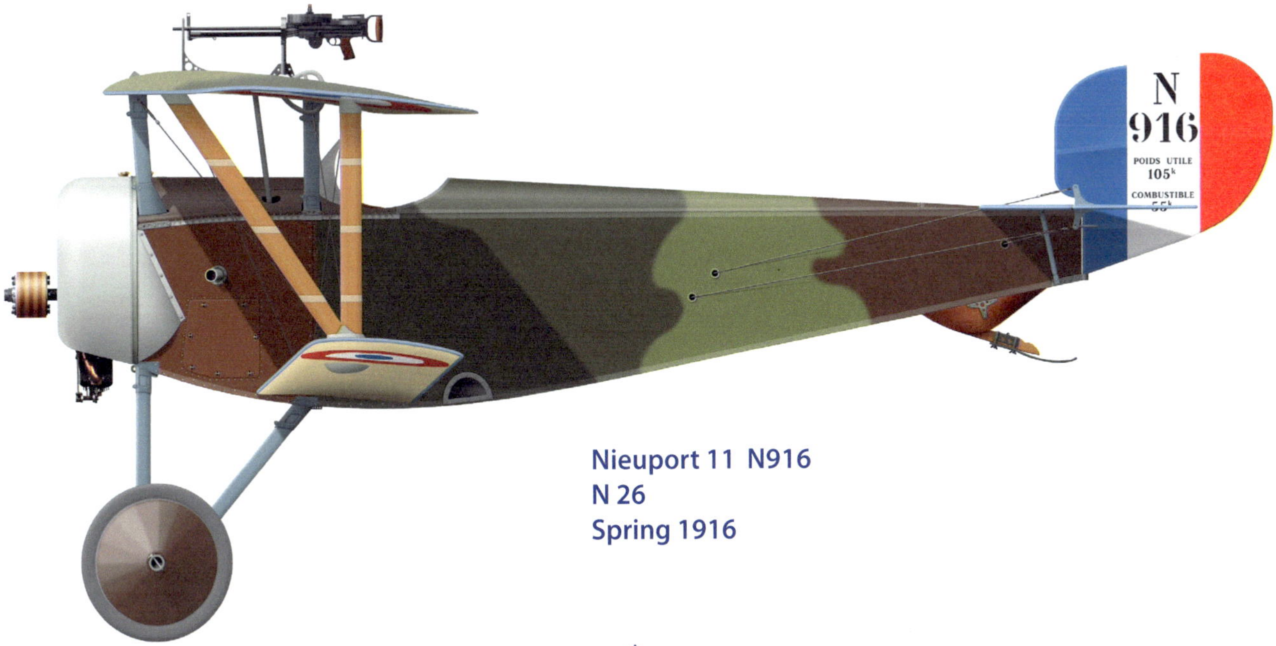

Nieuport 11 N916
N 26
Spring 1916

Nieuport 17 N1581
N 26
August–September 1916

Jean Charles Romatet
(1893–1975)
Lieutenant
N 38, Spa 76, Spa 165
7 victories

Spad 13 S10054
Spa 165
Late summer–autumn 1918

Jacques Raphaël Roques
(1897–1988)
Adjudant
Spa 48
5 victories

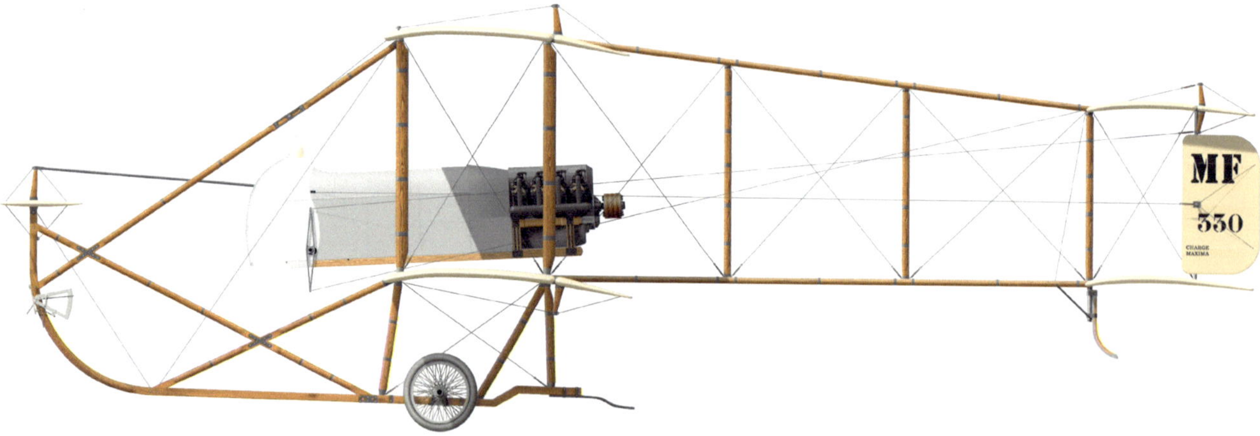

Maurice Farman 7 MF330
Military Flying School Chartres
March 1916

Nieuport 11 N1359
Military Flying School Pau
Summer 1916

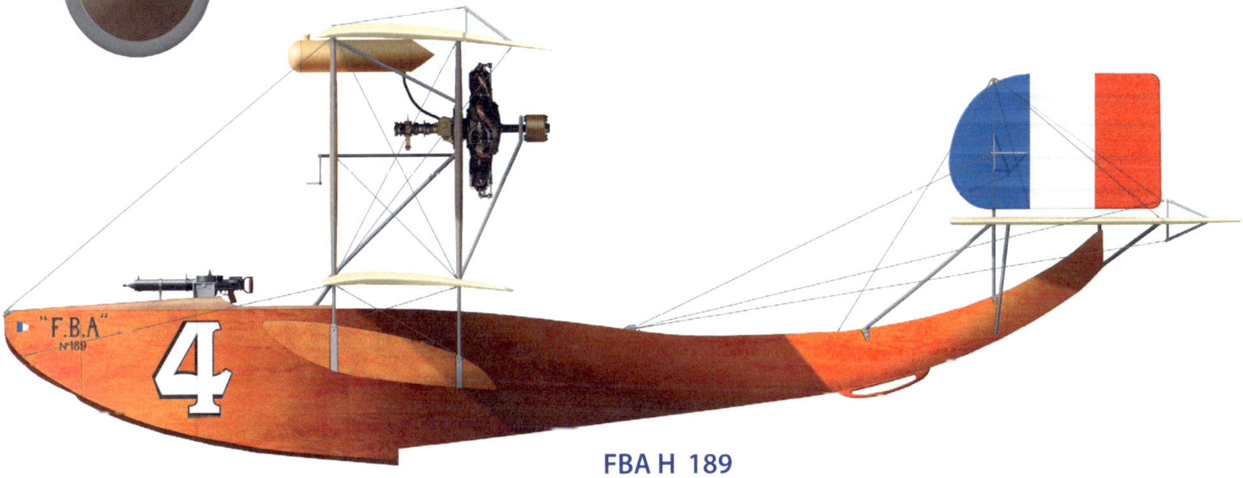

FBA H 189
Military Flying School Casot
Autumn 1916

Nieuport 17 N2039
Spa 48
January 1917

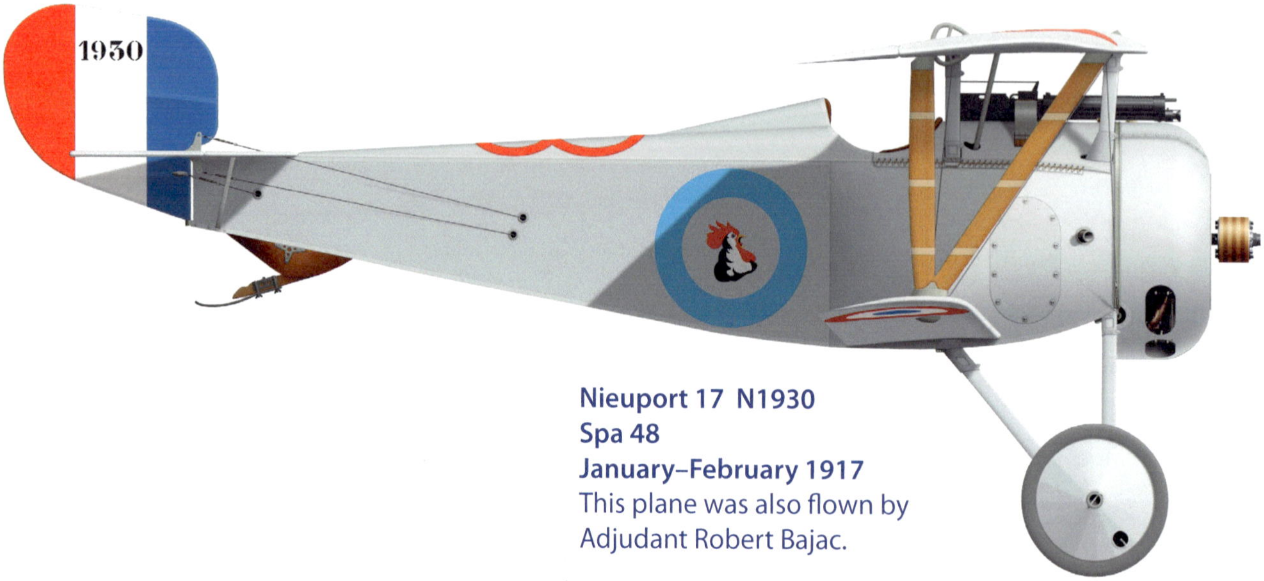

Nieuport 17 N1930
Spa 48
January–February 1917
This plane was also flown by Adjudant Robert Bajac.

Spad 7 *Lardon II* S1188
Spa 48
April 1917

Spad 7 S1213
Spa 48
Summer 1917

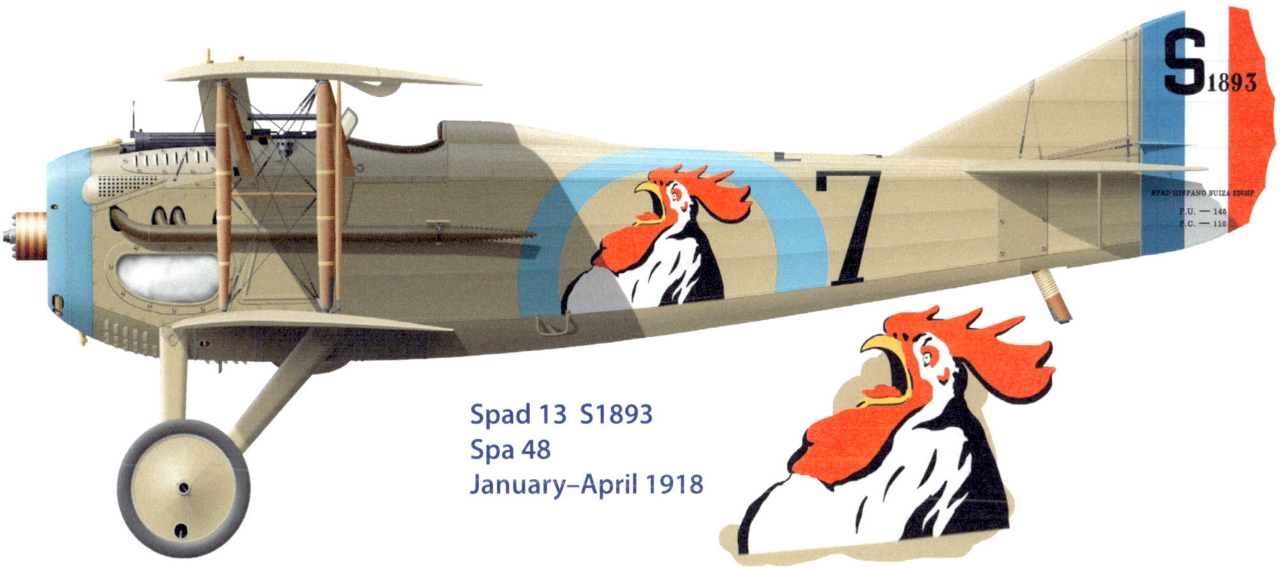

Spad 13 S1893
Spa 48
January–April 1918

Spad 13 S727
Spa 48
January–April 1918

Achille Justin Ernest Rousseaux
(1887–1928)
Adjudant
C 46, N 79, N 88
6 victories

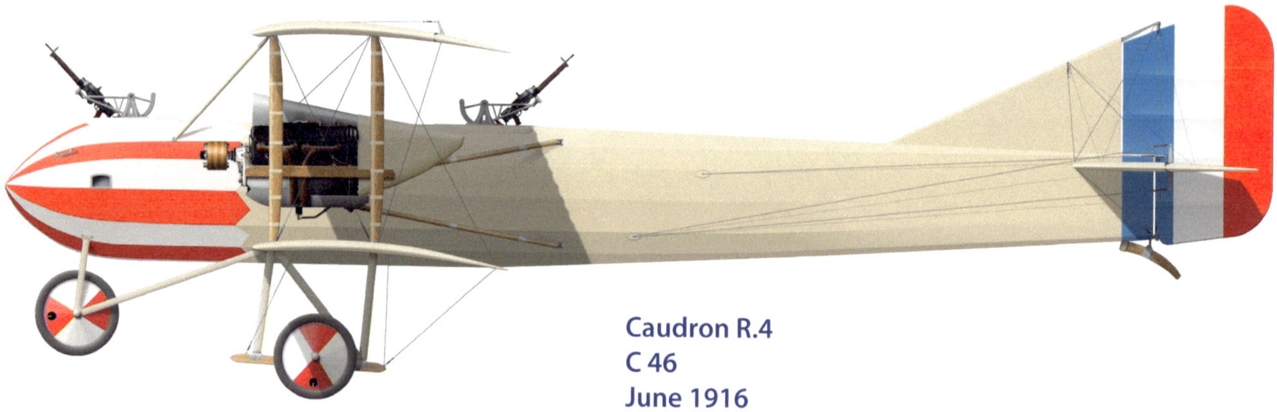

Caudron R.4
C 46
June 1916

Paul Marie Raphael Santelli
(1898–1928)
Maréchal des logis
Spa 81
7 victories

Spad 7
Spa 81
Winter 1918

Jean Marie Luc Gilbert Sardier
(1897–1976)
Lieutenant
N 77, Spa 77, Spa 48
15 victories

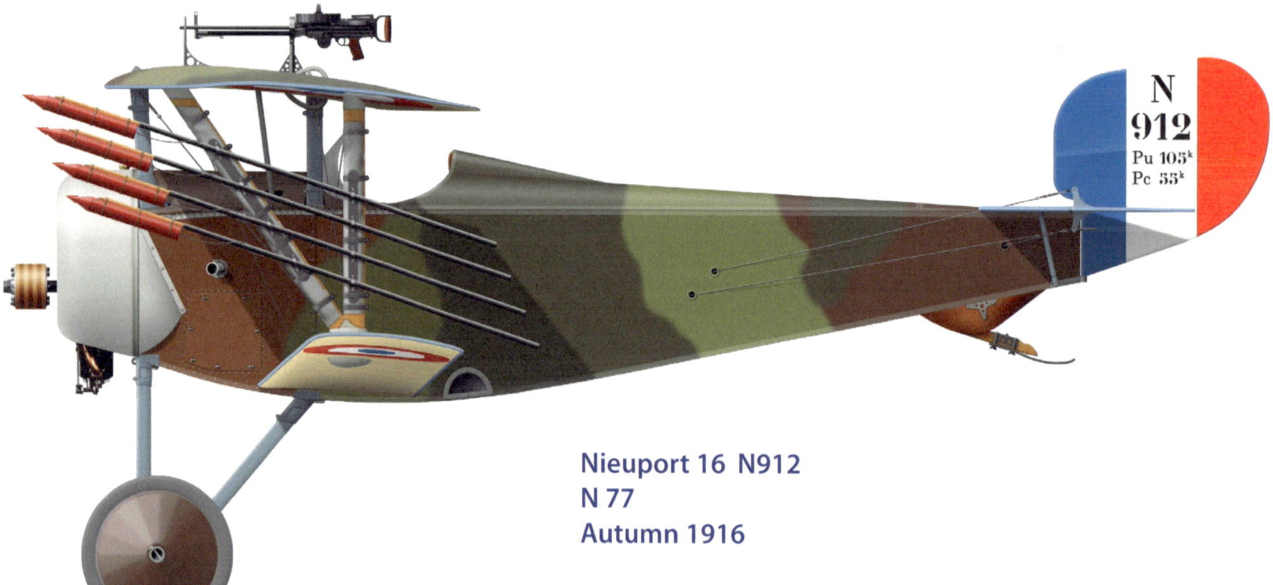

Nieuport 16 N912
N 77
Autumn 1916

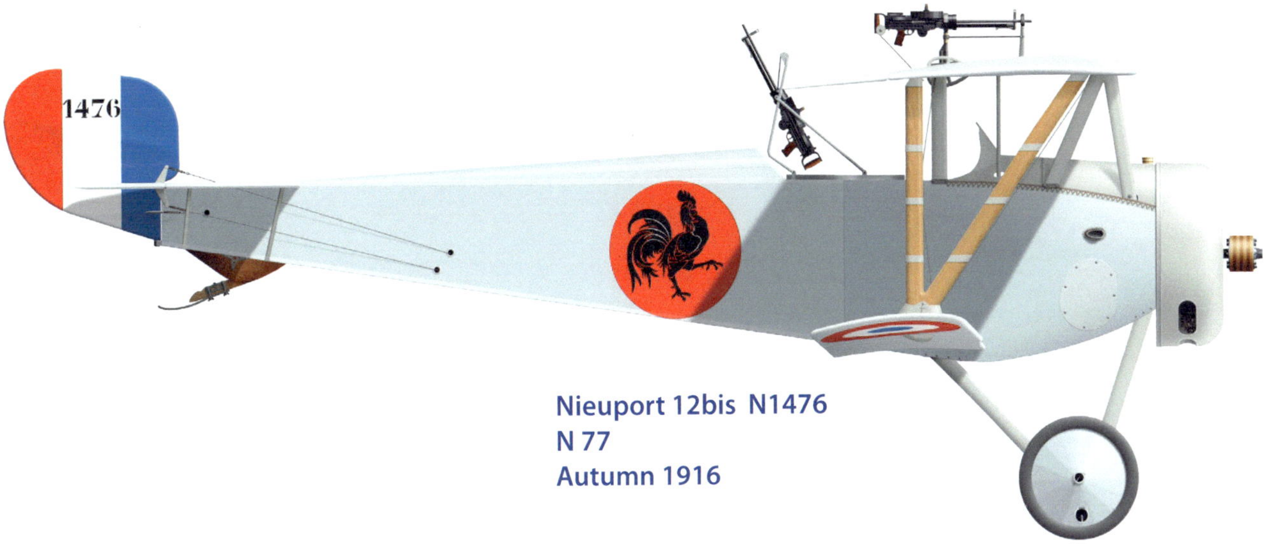

Nieuport 12bis N1476
N 77
Autumn 1916

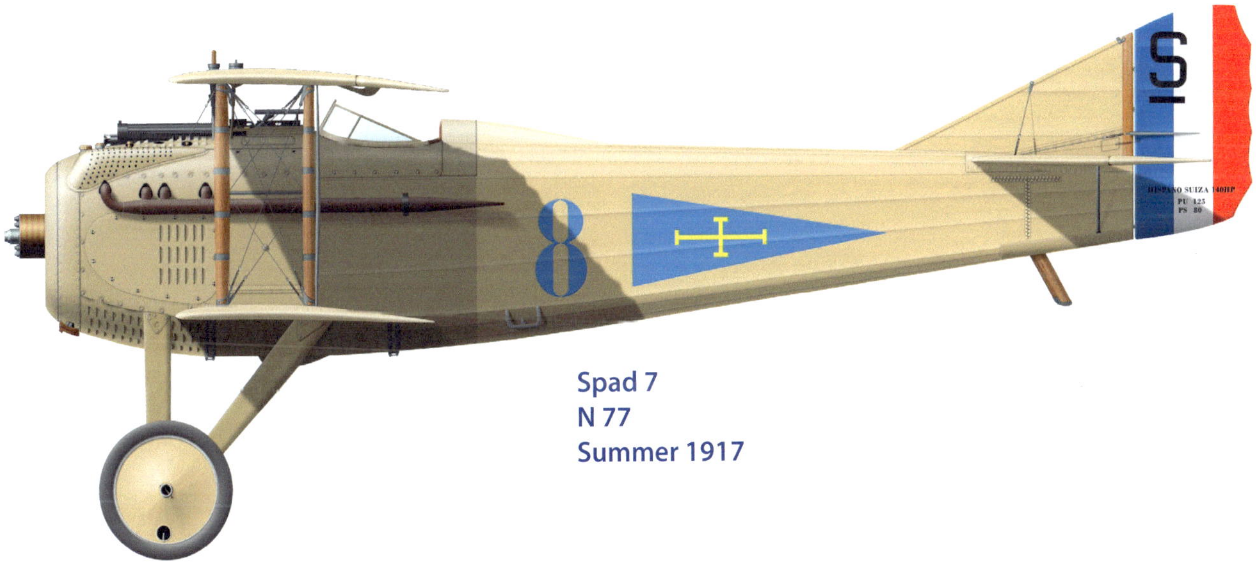

Spad 7
N 77
Summer 1917

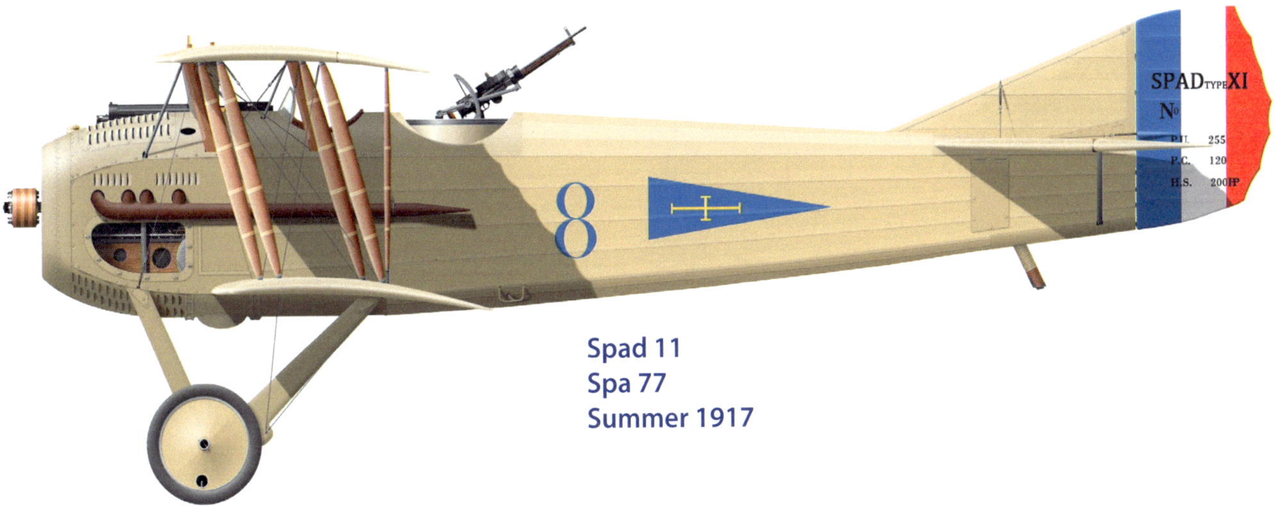

Spad 11
Spa 77
Summer 1917

Spad 13
Spa 77
May–June 1918

Spad 13
Spa 48
Summer 1918

Basile Baptiste Félicien Sauné
(1887–1918)
Sous-Lieutenant
MF 85, N 531
5 victories

Nieuport 17 N1197
Military Flying School Pau
Autumn 1916

Nieuport 23
N 532
May 1917

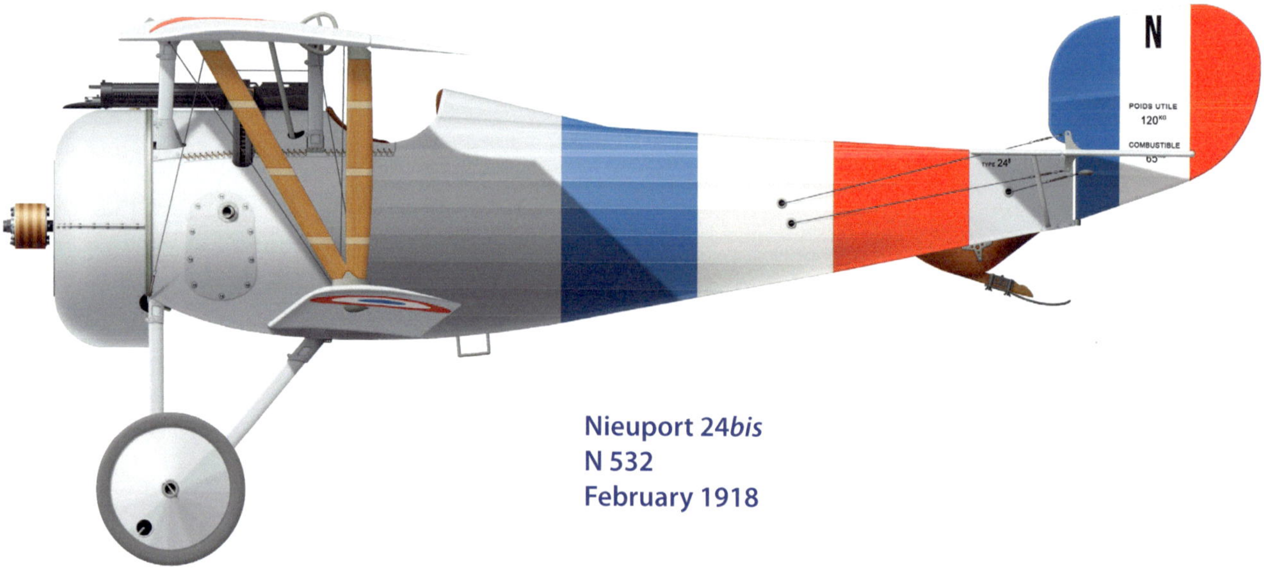

Nieuport 24*bis*
N 532
February 1918

Paul Jean Sauvage
(1897–1917)
Sergent
N38, N65
7 victories

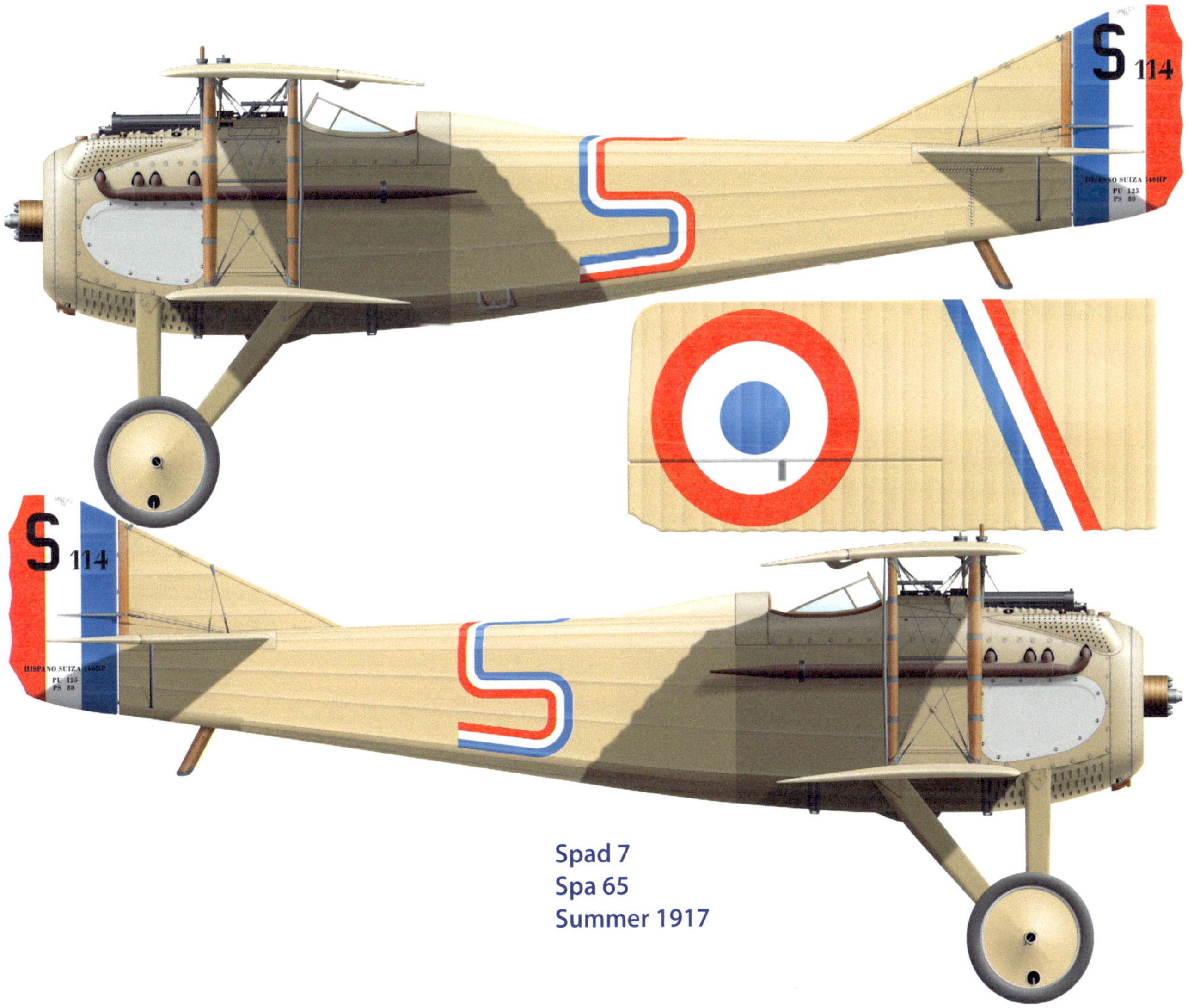

Spad 7
Spa 65
Summer 1917

Victor Louis Georges Sayaret
(1889–1980)
Adjudant-Chef
V 24, N 57, N 76
7 victoriess

Nieuport 17 N1365
N 57
August 1916

Joseph Marie Xavier de Sevin
(1894–1963)
Capitaine
N 12, Spa 26
12 victories

Nieuport 17 N1431
N 12
June–July 1916

Nieuport 17 N1431
N 12
Autumn 1916

Spad 7 S220
Spa 12
November–December 1917

Spad 13
Spa 26
September 1918

Constant Frederic Soulier
(1897–1933)
Marechal-des-Logis
N 26
6 victories

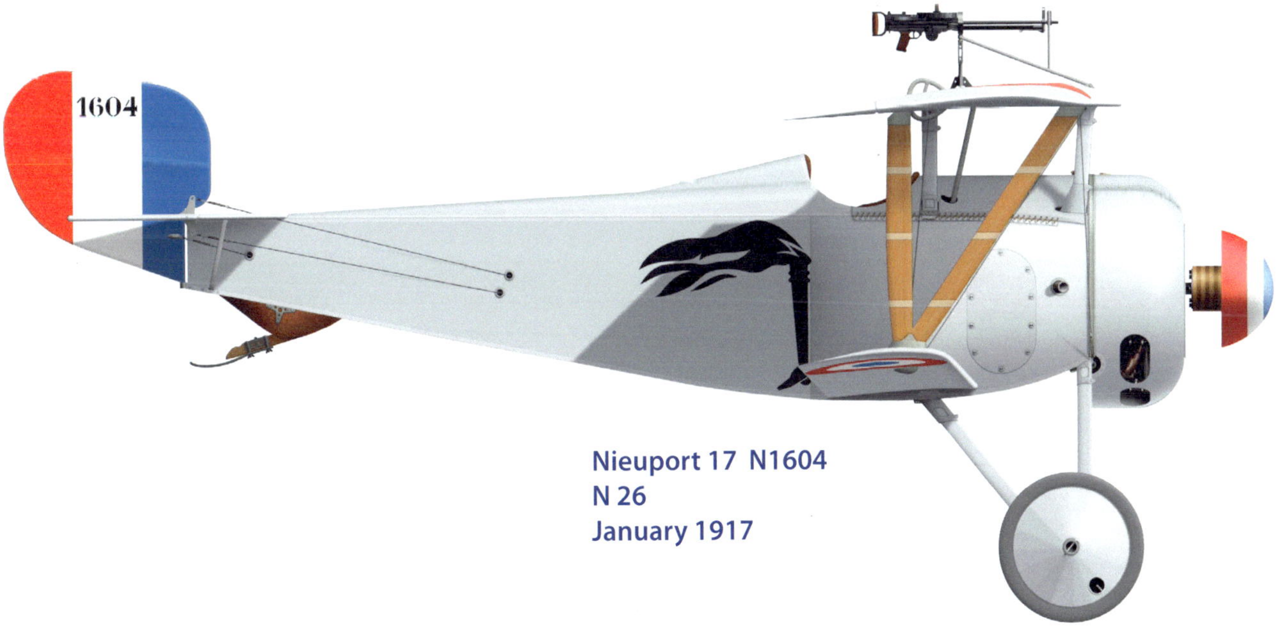

Nieuport 17 N1604
N 26
January 1917

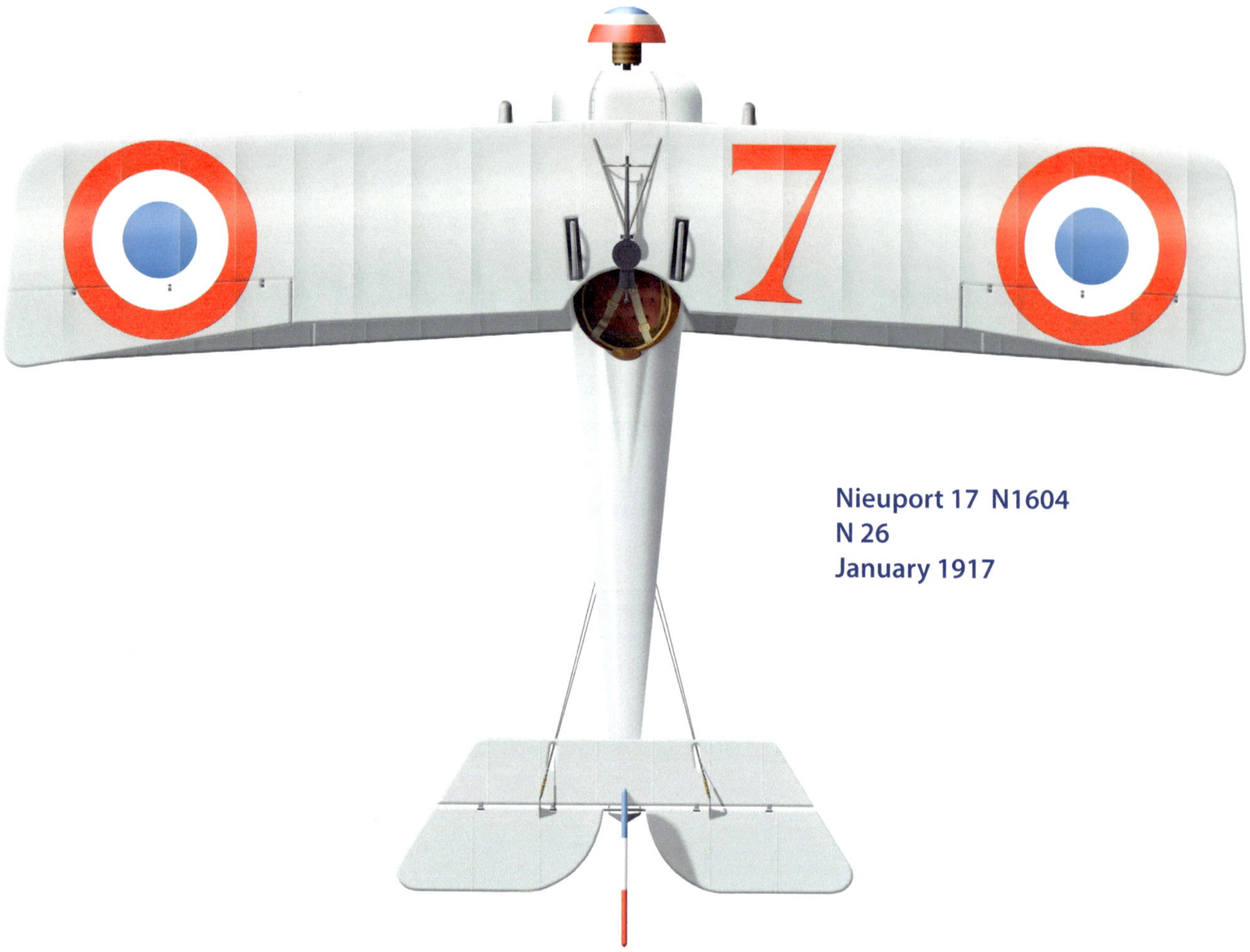

Nieuport 17 N1604
N 26
January 1917

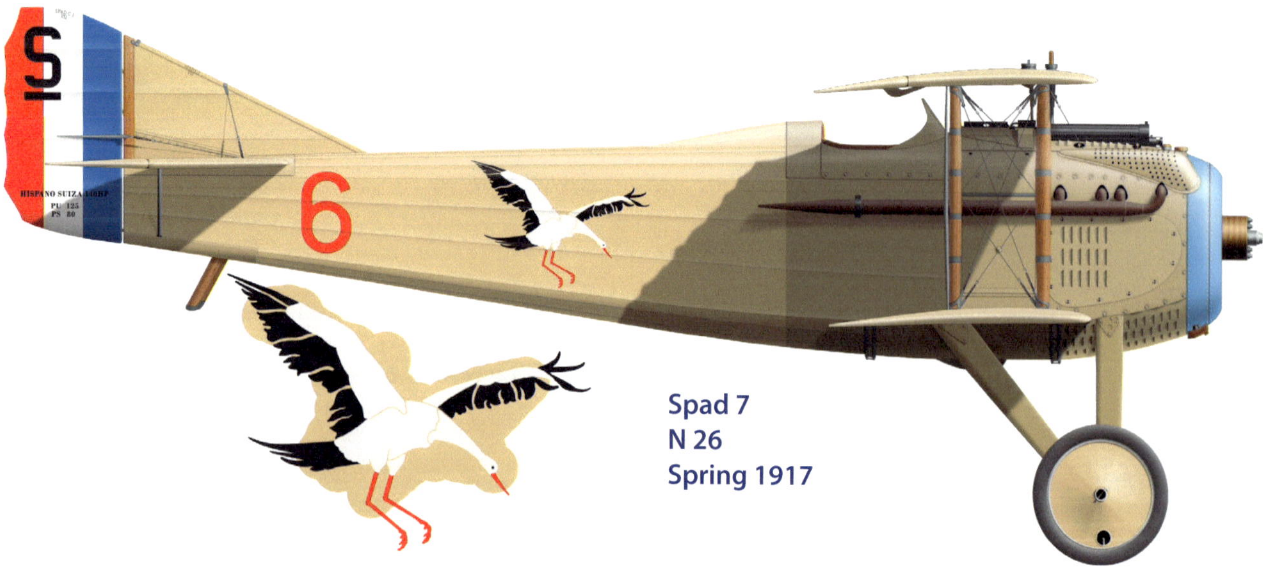

Spad 7
N 26
Spring 1917

Paul Albert Pierre Tarascon

(1882–1977)
Lieutenant
N31, N3, N62
12 victories

Nieuport 11 *ZIGOMAR* N1159
N 62
May–June 1916

Nieuport 17 N1427
N 62
16 July–late July 1916

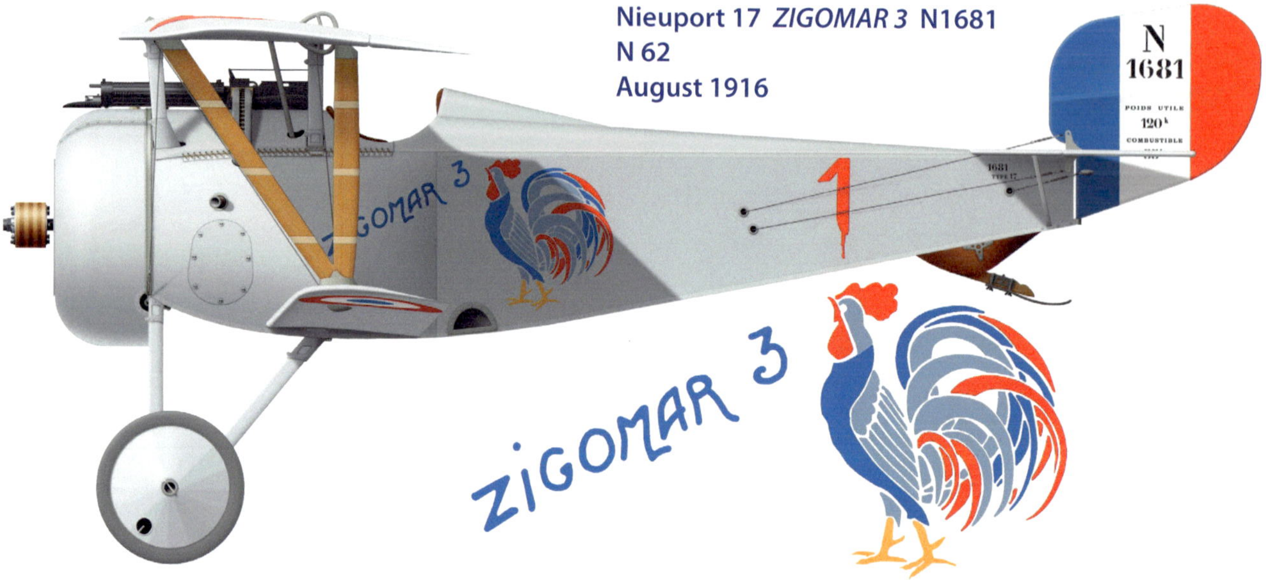

Nieuport 17 *ZIGOMAR 3* N1681
N 62
August 1916

Nieuport 23 *ZIGOMAR 4* N1662
N 62
January 1917

Nieuport 24*bis* *ZIGOMAR 5* N3588
N 62
April 1917

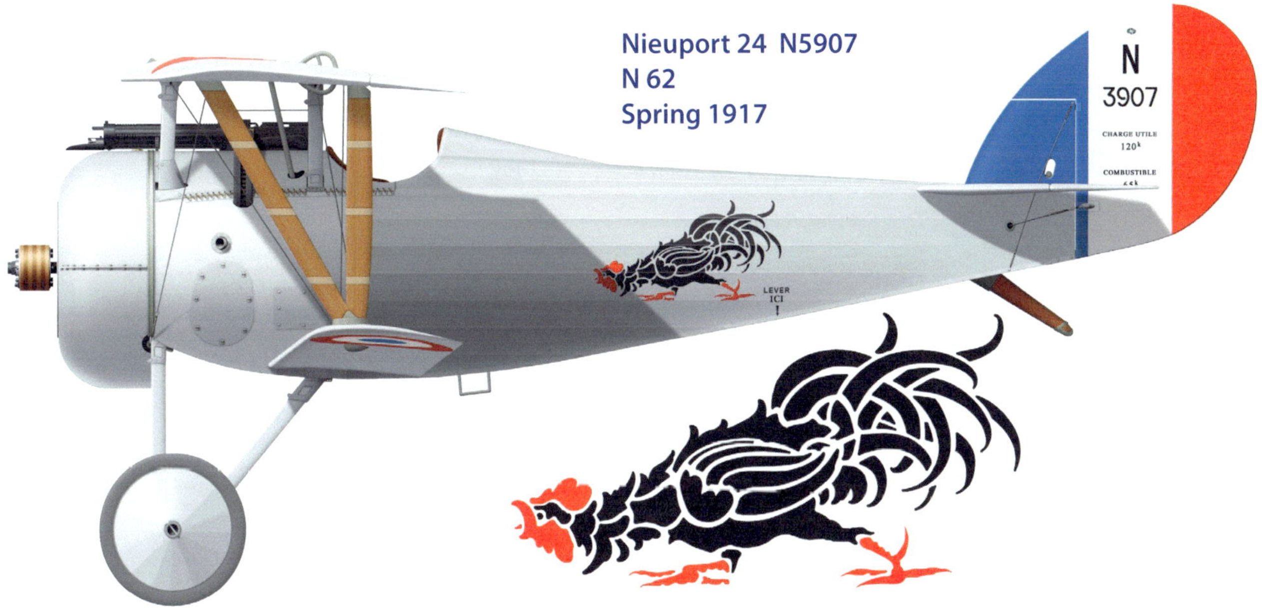

Nieuport 24 N5907
N 62
Spring 1917

Nieuport 28 N6187
August 1917
Demonstration flights for Americans.

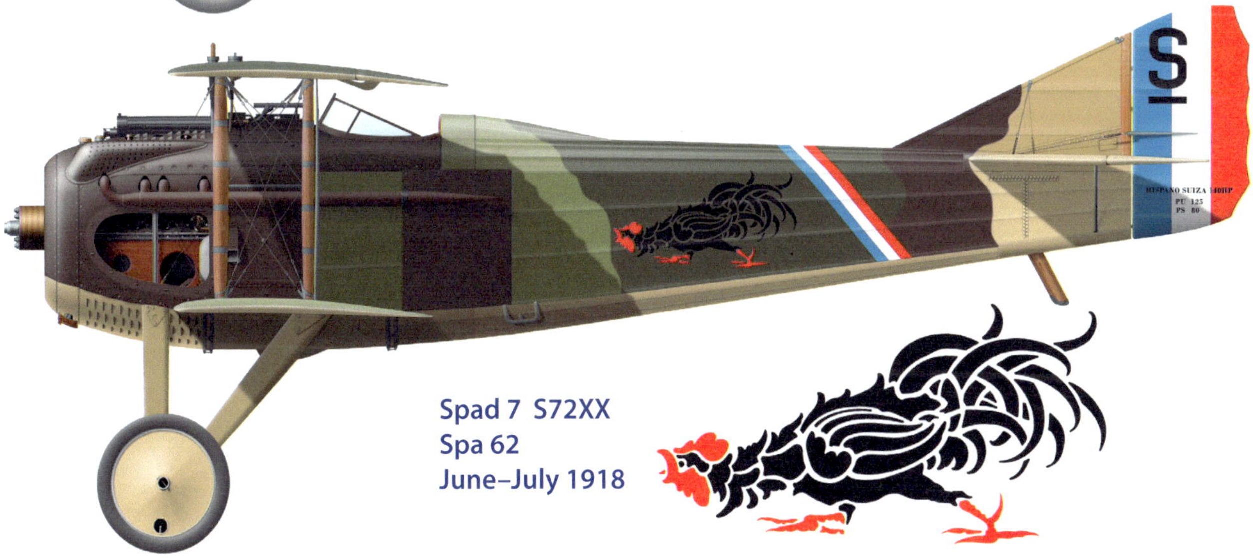

Spad 7 S72XX
Spa 62
June–July 1918

Mathieu Marie Joseph Antoine Tenant de la Tour (1883–1917)

Capitaine
MS 15, N 67, N 57, N 3, Spa 26
9 victories

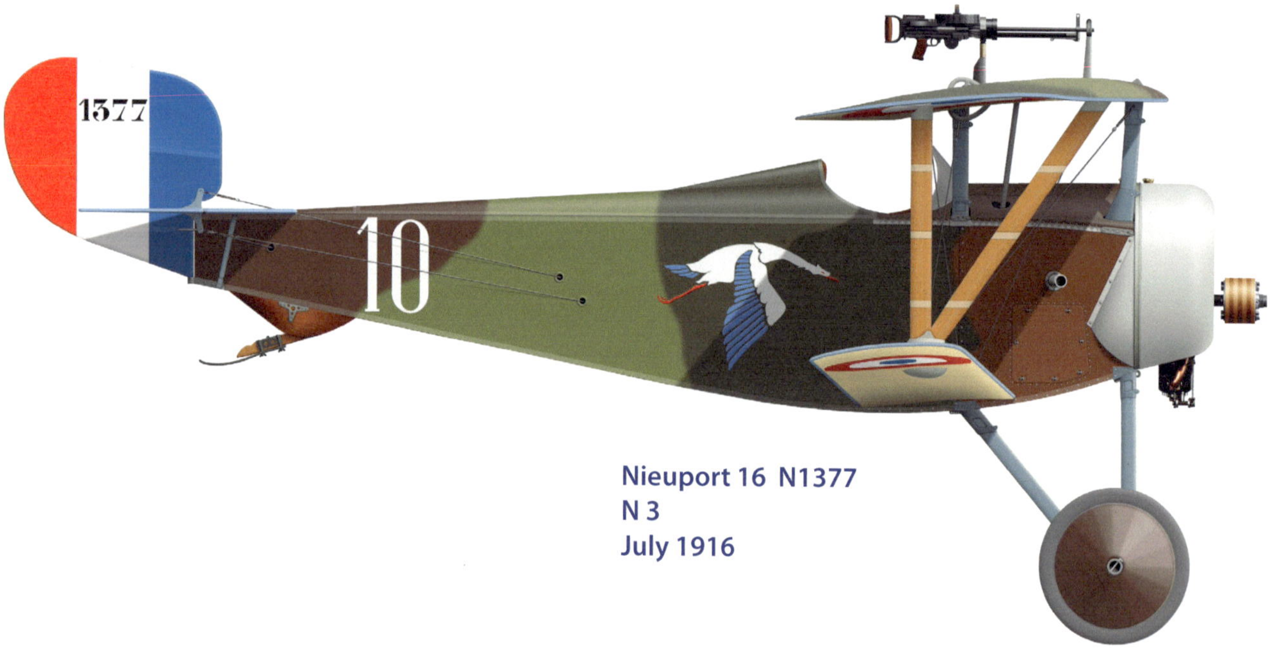

Nieuport 16 N1377
N 3
July 1916

Nieuport 17 N1560
N 3
August–September 1916

Nieuport 17 N1560
N 3
September 1916

Spad 7 S184
N 3
Late December 1916

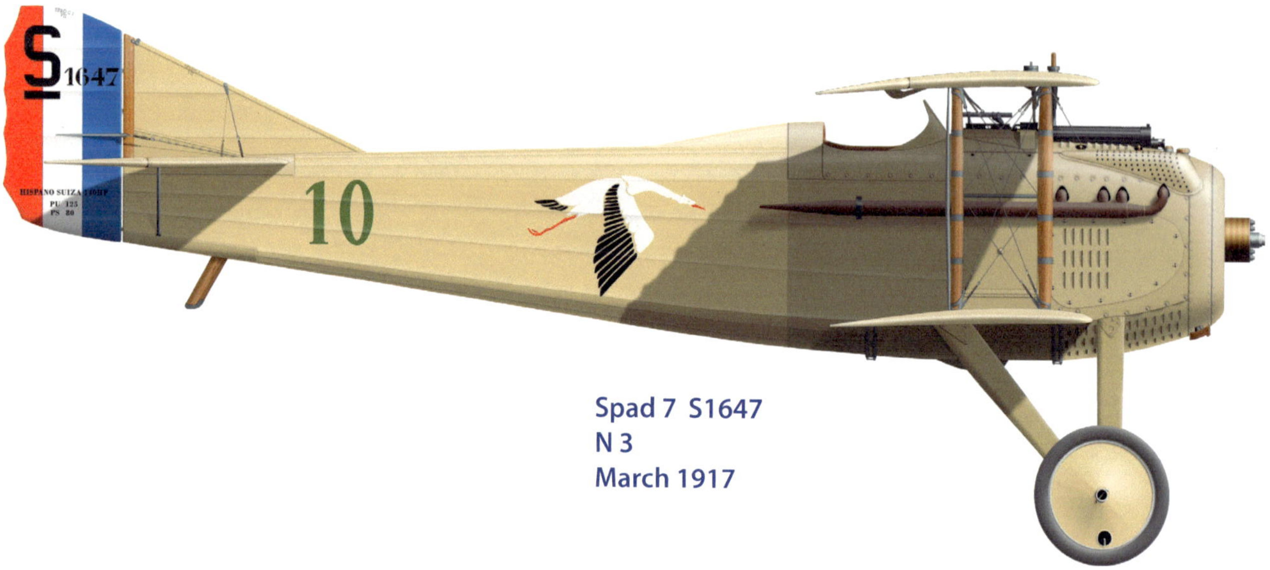

Spad 7 S1647
N 3
March 1917

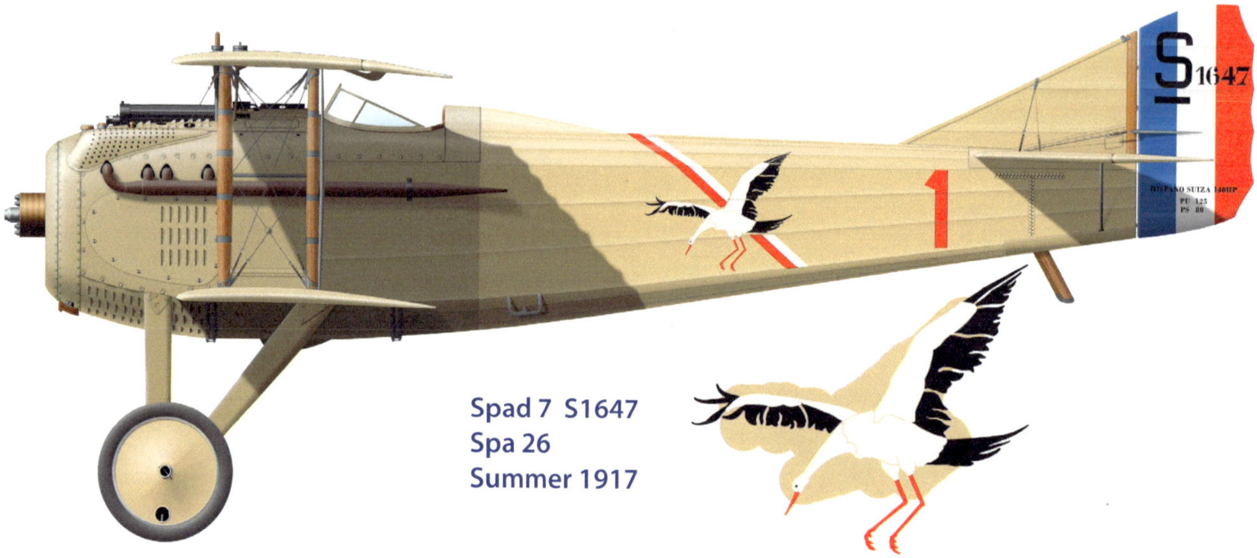

Spad 7 S1647
Spa 26
Summer 1917

Armand Jean Galliot Joseph de Turenne
(1891–1980)
Capitaine
GB 1, N 48, Spa 12, GC 24
15 victories

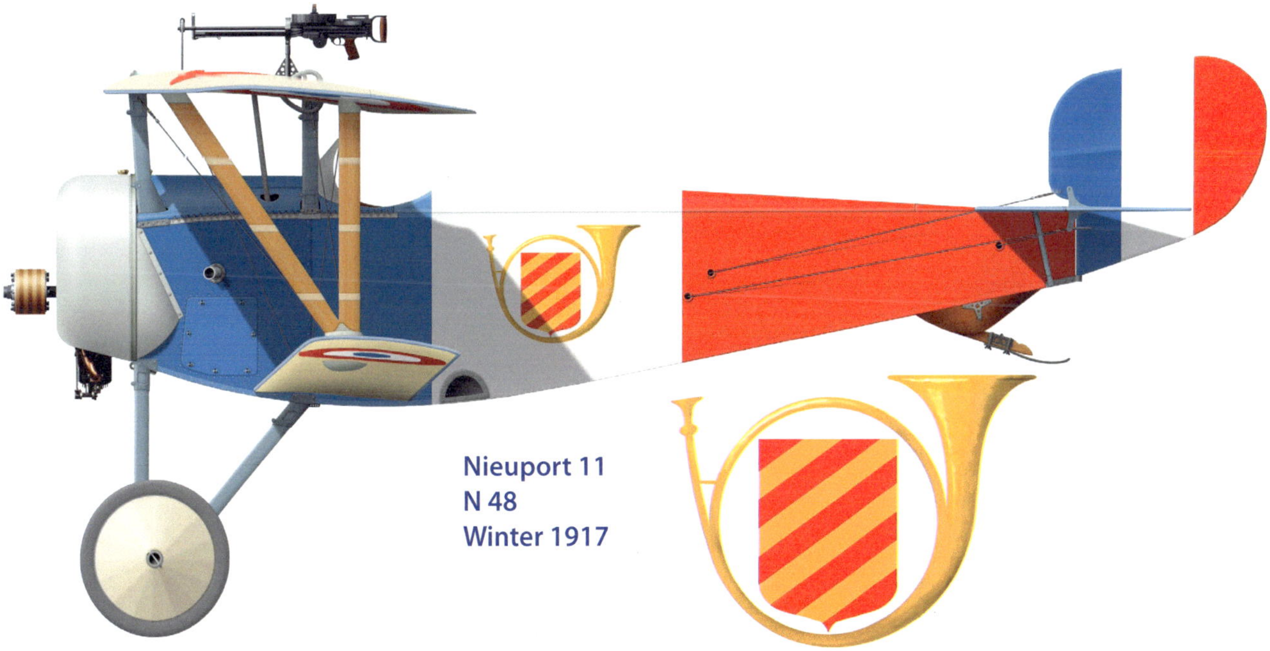

Nieuport 11
N 48
Winter 1917

Armand Turenne

Nieuport 17 N2186
N 48
Spring 1917

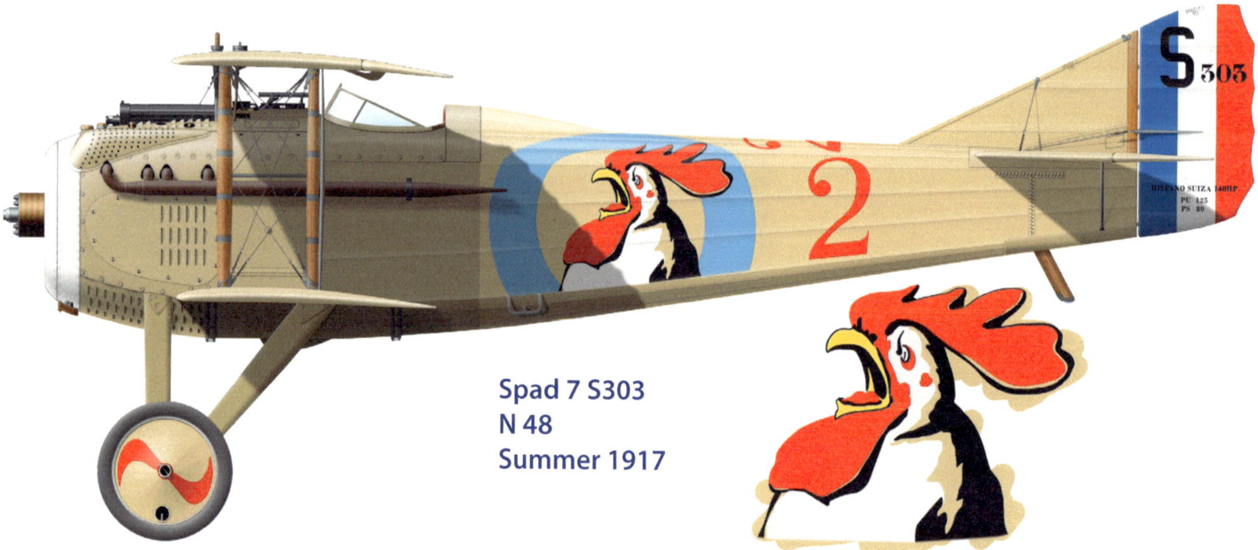

Spad 7 S303
N 48
Summer 1917

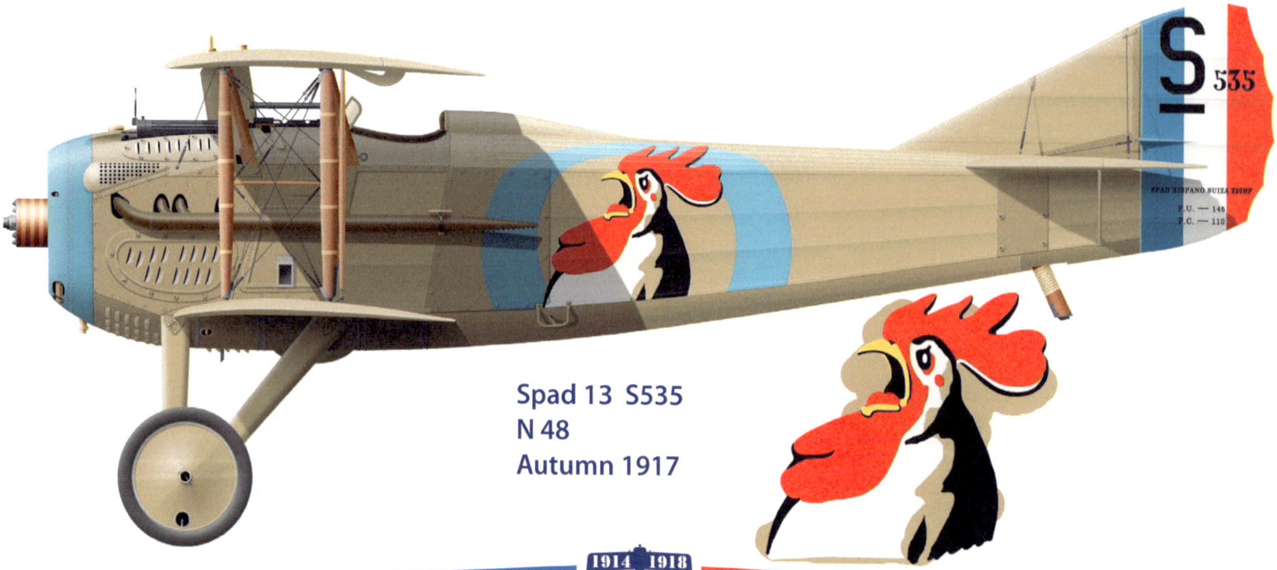

Spad 13 S535
N 48
Autumn 1917

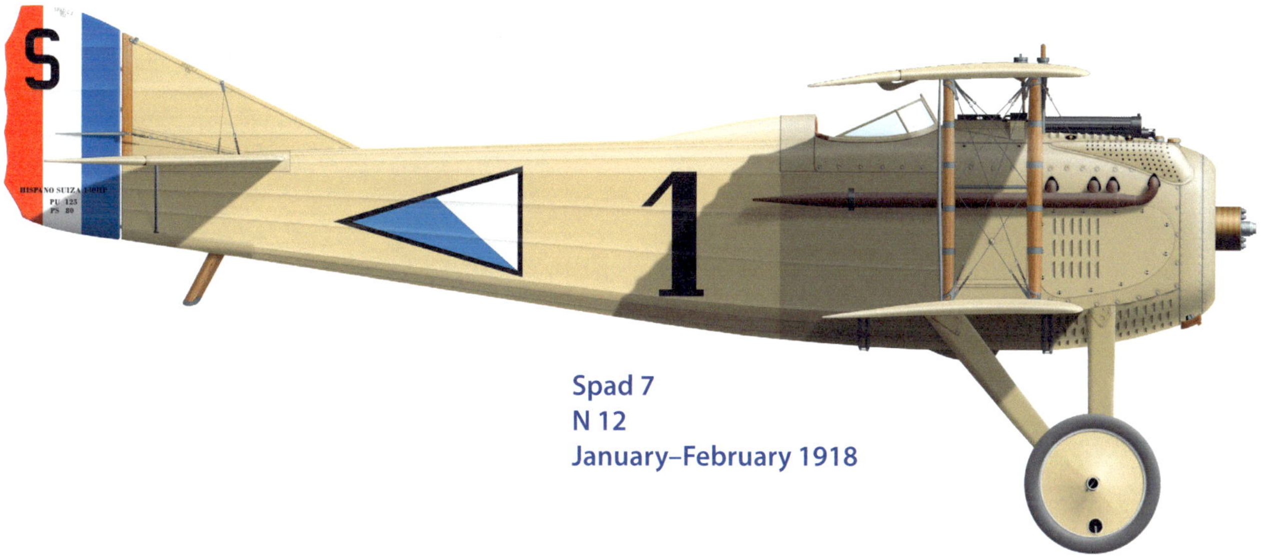

Spad 7
N 12
January–February 1918

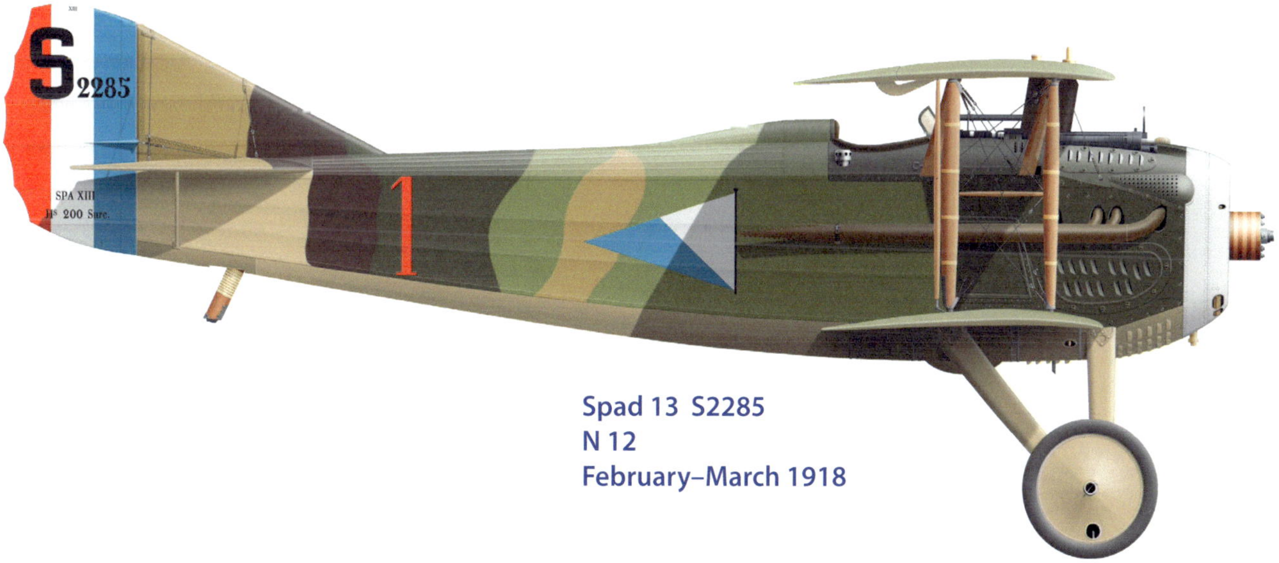

Spad 13 S2285
N 12
February–March 1918

Del Antoine Gaston Vial
(1891–1982)
Sous-Lieutenant
BL9, MS3, MF29, F123, N315, Sop29, Spa152, Spa315
5 victories

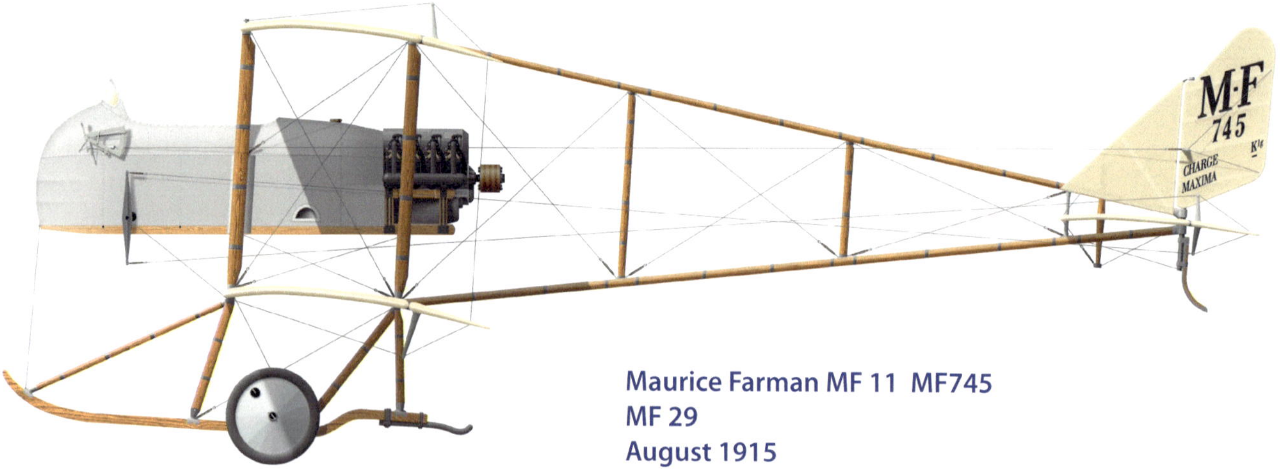

Maurice Farman MF 11 MF745
MF 29
August 1915

Marcel Pierre Viallet
(1887–1925)
Sous-Lieutenant
C53, N67
9 victories

Morane-Saulnier AC 865
N 67
Summer 1916

Pierre Augustin François Violet-Marty
(1894–1916)
Adjudant
MF 55, N 57
5 victories

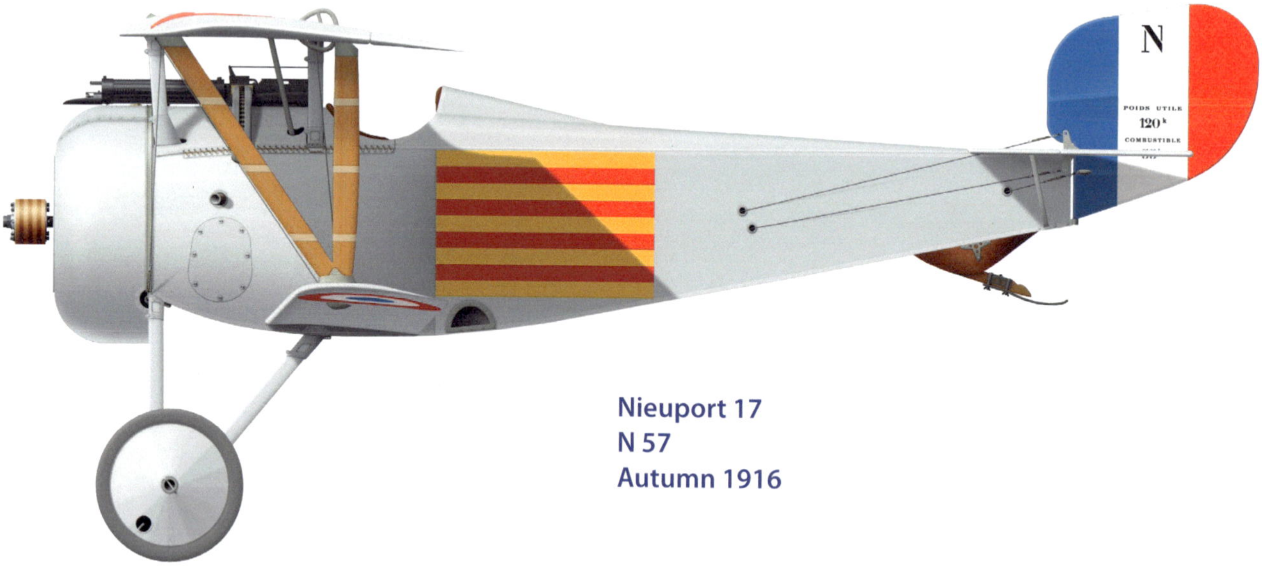

Nieuport 17
N 57
Autumn 1916

Marie Gaston Fulerand Leon Vitalis
(1890–1941)
Adjudant-Chef
C 46, N 67
6 victories

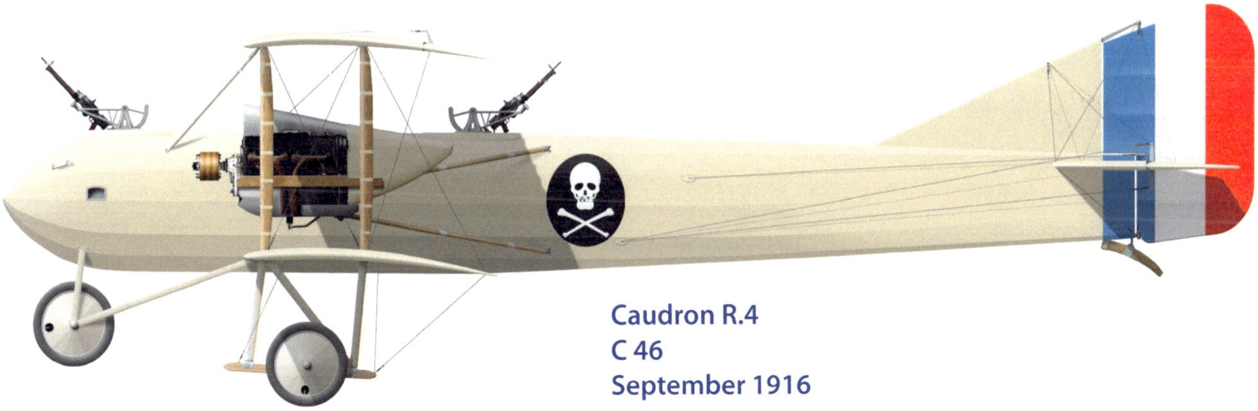

Caudron R.4
C 46
September 1916

Joseph Vuillemin
(1883–1963)
Chef d'Escadron (Major)
C 11, C 39, Esc 12, GB 5
7 victories

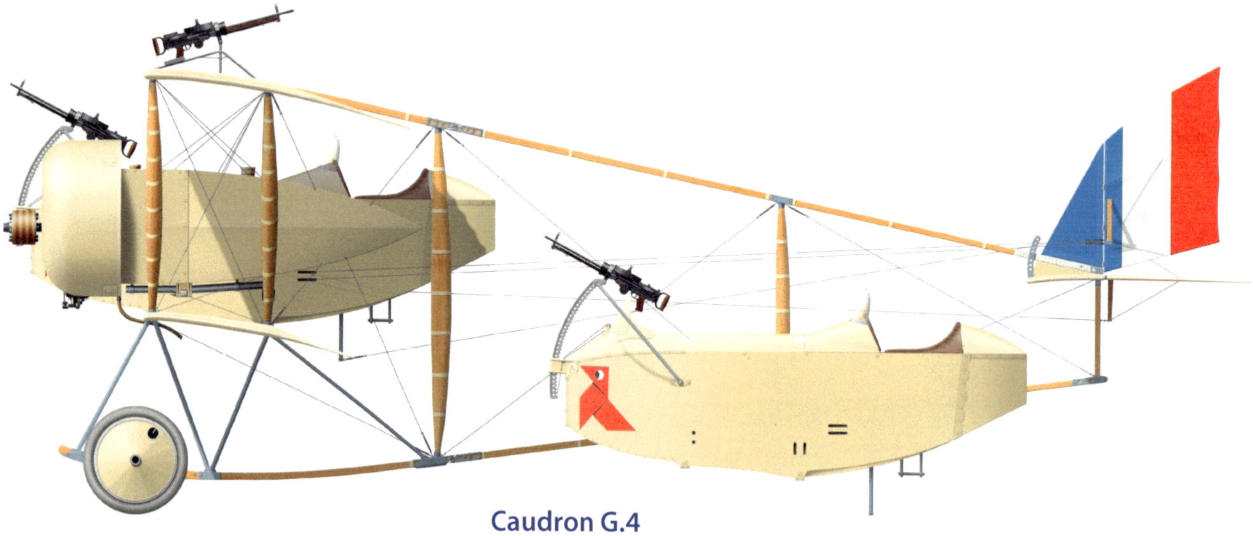

Caudron G.4
C 11
March 1916

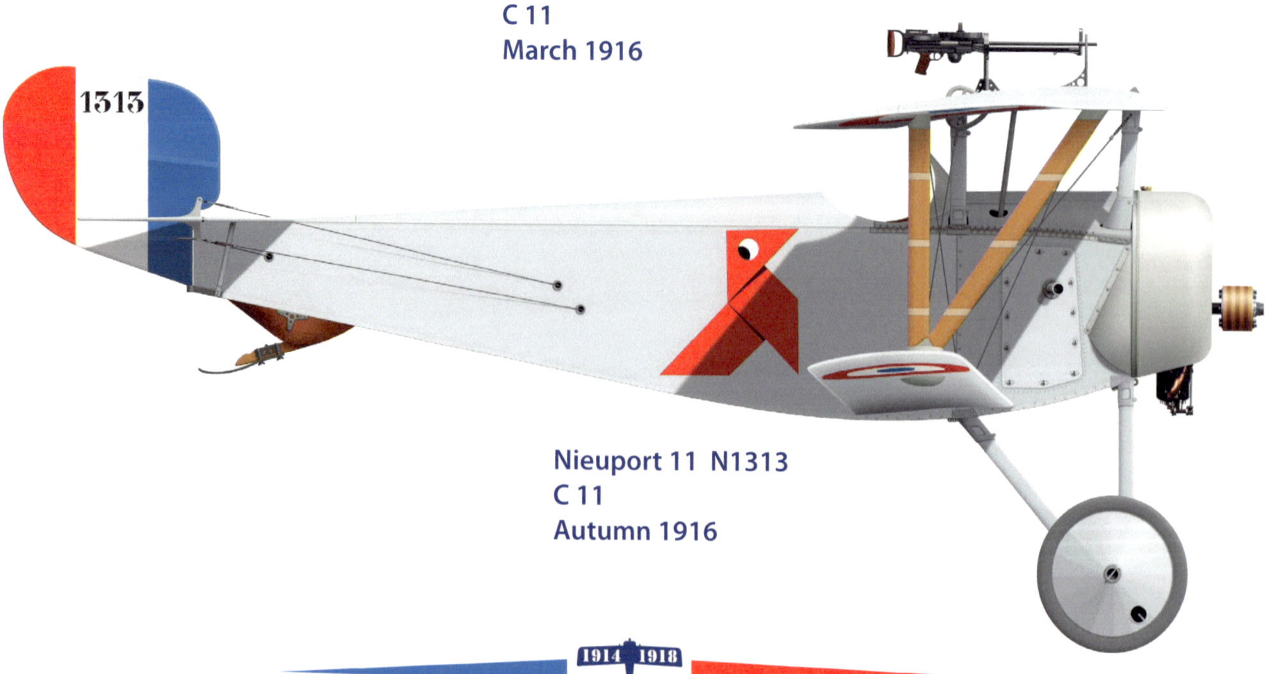

Nieuport 11 N1313
C 11
Autumn 1916

Paul Yvan Robert Waddington
(1893–1986)
Sous-Lieutenant
N 12, N 67, Spa 31, Spa 154
12 victories

Spad 13 S8213
Spa 154
Summer 1918

Eugène Weismann
(1896–1973)
Sous-Lieutenant
BR 132
7 victories

Breguet 14 B2
Br 132
August 1918
This plane was flown by *Capt.* J.F. Jannekeyn (obs.) and *Sous-Lt.* Weismann (pilot).

Breguet 14 B2 4231
Br 132
September 1918

Pierre Armand Wertheim (Wertheimer) (1888–1971)

Lieutenant
Spa84
5 victories

Spad 13
Spa 84
Autumn 1918
Recontruction

Sources & References

Books

Ezra Bowen, *Knights of the Air*, Time-Life Books, 1980
Marc Chassard, Nieuport 17 Analyse détaillée des premières séries, Association FAREWO, 2018.
Marc Chassard, *René Dorme et Joseph Guiguet. La Guerre aérienne de deux As*, ÉDITIONS AUX ARTS, 2002.
James J. Davilla, Artur M. Soltan, *French Aircraft of the First World War*, Flying Machines Press, 1997.
Alan Durkota, Thomas Darcey, Victor Kulikov, *The Imperial Russian Air Service. Famous Pilots and Aircraft of World War I*, Flying Machines Press, 1995.
A. van Haute, *Pictorial History of the French Air Force*, Ian Allan, London, 1974.
Jack Herris, *SPAD Two-seat Fighters of World War I*, Flying Machines Press, 2005.
Stephen Longstreet, *The Canvas Falcons*, Barnes & Noble, 1970
Norman L.R. Franks, Frank W. Bailey, *Over the Front*, Grub Street, 1992.
Norman L.R. Franks, Frank W. Bailey, Russel Guest, *Above the Lines*, Grub Street, 1993.
Norman L.R. Franks, *Who Downed the Aces in WWI?*, Grub Street, 1996.
Norman L.R. Franks, Frank W. Bailey, *The Storks*, Grub Street, 1998.
Norman Franks, *Nieuport Aces of World War 1*, Osprey Publishing, 2000.
Norman L.R. Franks, Christophe Cony, *The French Air Service War Chronology 1914–1918*, Grub Street, 2001.
Vital Ferry, *French Aviation during the First World War, Histoire and Collections*, 2015
Réne Fonk, *Mes Combats*, (Ace of Aces), Doubleday & Co, 1967.
Jon Guttman, *SPAD VII Aces of World War 1*, Osprey Publishing, 2001.
Jon Guttman, *SPAD XII/XIII Aces of World War 1*, Osprey Publishing, 2002.
Jon Guttman, *SPA 124 Lafayette Escadrille*, Osprey Publishing, 2004.
Jon Guttman, *Groupe de Combat 12, 'Les Cigognes'*, Osprey Publishing, 2004.
Jon Guttman, *Balloon-Busting Aces of World War 1*, Osprey Publishing, 2005.
Jon Guttman, *Pusher Aces of World War 1*, Osprey Publishing, 2009.
Jon Guttman, *Resonnaisance and Bomber Aces of World War 1*, Osprey Publishing, 2015.
Jon Guttman, *Grim Reapers: French Escadrille 94 in World War I*, Aeronaut Books, 2016.
Jules Roy, Guynemer, L'Ange de La Mort, Albin Michel, 1986.
Jan Sumner, *The Kings of the Air: French Aces and Airmen of the Great War*, Pen and Sword Digital, 2015.
Jim Wilberg, Jean Navarre: *France's Sentinel of Verdun and the First French Fighter Pilots*, Aeronaut Books, 2010
Georges Thomas, *Guynemer, l'as des as*, Imp. de Vaugirard, 1917 (Gallica)

Periodicals

Aérojournal – No. 18, 26, 41, 47, 48, 49
Avions – Vols 1 to 235
Batailles Aeriennes – No. 74, 77, 81
Bulletin mensuel / Ligue aéronautique de France – 1917
Cross & Cockade (International) – Vols 1 to 44
Cross & Cockade (US) – Vols 1 to 26
HMP – No. 5, 6, 12/1992, 7/ 1995, 11/2003
La Vie aérienne : revue hebdomadaire paraissant le jeudi – 02/09/1919 (Gallica)
La Guerre aérienne illustrée – 1916, 1917, 1918, 1919 (Gallica)
Le Fana de l'Aviation – Vols 6 to 553
Le Monde illustré 1921 (Gallica)
Lotnictwo z szachownica – No. 1/ 2002
Modelář – No. 4/2004
Over the Front – Vols 1 to 31
Replic – No.9, 12, 43, 86, 194
Revi – No. 4, 50, 51
Squadron Signal No. 93 SPAD Fighters
Windsock International – Vols 1 to 31
Windsock Datafile – No. 8, 16, 20, 32, 36, 58, 68, 80, 94, 96, 108, 109, 135, 161, 167
Windsock Mini Datafile – 3, 5, 6
Windsock Datafile Special – Nieuport Fighters (Vols 1, 2), Breguet 14
Wing Masters – No. 61

Internet Resources

www.albindenis.free.fr
www.airwar.ru
www.as14-18.net
www.asoublies14-18.cosadgip.com
www.aviaww1.forum24.ru
www.bibert.fr
www.cieldegloire.com
www.deullin.fr
www.fandavion.free.fr
www.gallica.bnf.fr
www.memoiredeshommes.sga.defense.gouv.fr
www.mincbergr.net
www.navarre-jean.com
www.pages14-18.mesdiscussions.net
www.rene.fonck.free.fr
www.scalemodels.ru
www.theaerodrome.com
www.traditions-air.fr

Made in United States
Cleveland, OH
06 November 2026

47771494R10088